MAKERPHONES

The Art and Science of Headphone Design

Designing and building your own headphones — the real engineering, explained plainly.

Jamey Warren

The Art and Science of Headphone Design
First edition · 2026

Published by Warren Labs.
ISBN 979-8-9968299-0-3 (paperback)
ISBN 979-8-9968299-1-0 (ebook)
Set in Schibsted Grotesk, Source Serif 4, and JetBrains Mono.
This edition is generated from the live manual at makerphones.com via the project's to-book pipeline. For interactive figures and the 3D parts viewer, visit the website.

For Piper and Jessica.

For my parents, Curt and Margie, and my sister, Jennifer.

And for all my family, near and far.

Contents

APPENDICES

Figures

How to use this book

This is a reference manual you can read two ways. Read it front to back and it builds an argument: how headphones make sound, what the parts do, how to design and build an enclosure, how to measure what you made, and how to tune it. Or treat it as a reference — each chapter names what you should have read first, so you can come in from any direction and find your footing.

The six numbered parts climb from beginner to advanced. After them, the build guides walk through real headphones end to end, and the appendices collect a glossary, suppliers, resources, sources, troubleshooting, and a standing note on listening safely. Figures are numbered by chapter — Fig. 4.2 is the second figure in Chapter 4 — and gathered in the list of figures up front.

Nothing here is theory for its own sake. Every claim that can be measured is, and the curve is shown. Where a value is a judgment call, it says so. The aim is a book you can trust at the bench: specific, unit-bearing, and honest about what's still uncertain.

A few things live only on the website — most of all the interactive 3D parts viewer. Where the page can't carry them, the book points you to makerphones.com.

Acknowledgments

This book carries a lot of other people's teaching.

I owe my understanding of headphones, specifically, to **Tyll Hertsens** of HeadRoom. So much of what's in these pages traces back to him. Thank you, Tyll.

My real audio-engineering education came from **Michael and Eben Grace**. Seven years at Grace Design, working directly under the two brothers, taught me more than any classroom could — I've always called it my University of Rock & Roll. Before that, the Conservatory of Recording Arts & Sciences gave me the formal grounding to build on.

And to the others — mentors, colleagues, and friends in this small field — whose names I'll keep adding as I remember them: thank you.

PART

1

Fundamentals

CHAPTER 1

How Headphones Create Sound

Learn how headphones create sound: voice coil, magnet, and diaphragm working together. Essential knowledge for building your own.

8 min read makerphones.com/learn/how-headphones-create-sound

ⓘ NOTE

This hobby puts drivers an inch from your eardrums. Before you build or test anything, take two minutes with Listening Safely.

You know how when you're learning something new, there's always that one person who can explain it in a way that just *clicks*? That's what I'm going for here. After 25+ years working with headphones, from manufacturing at Grace Design to running HeadRoom, I've explained this stuff hundreds of times.

The Basic Magic Trick

At their core, headphones do something pretty remarkable: they turn electricity into sound waves that your brain interprets as music. Understanding *how* they do this isn't trivia. When you design your own, this is the stuff every design decision traces back to.

The Three Parts That Matter

A dynamic driver, the type used in the vast majority of headphones, has three main components working together:

The voice coil is a coil of very fine wire wound in a precise pattern. When electrical current flows through this coil, it creates a magnetic field. Remember making electromagnets in school with a nail and some wire? Same principle, just much more precise.

The permanent magnet creates a fixed magnetic field. This is usually a strong neodymium or ferrite magnet arranged in a specific structure. The voice coil sits in the gap of this magnetic field.

The diaphragm is the part that actually moves air, and moving air is what creates sound. It's attached to the voice coil, so when the voice coil moves, the diaphragm moves with it.

Here's how it works: Your music player sends an electrical signal to the voice coil. This signal is constantly changing. That's your music, converted to electricity. As the current changes, the electromagnetic field from the voice coil interacts with the permanent magnet's field. Sometimes they attract, sometimes they repel. This makes the voice coil move back and forth rapidly.

Since the diaphragm is attached to the voice coil, it moves too. And when the diaphragm moves, it pushes and pulls on the air in front of it, creating pressure waves. Those pressure waves travel through the air, into your ear canal, vibrate your eardrum, and your brain interprets them as sound.

Pretty elegant, right?

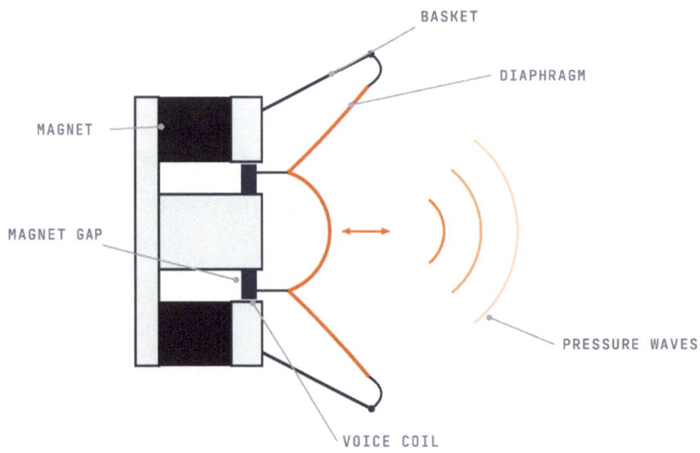

Fig. 1.1 — The dynamic driver in cross-section. The moving parts are in orange.

Why Headphones Aren't Just Tiny Speakers

I hear this one all the time: "Headphones are just little speakers, right?"

Not quite. And the difference matters once you start designing.

With speakers in a room, the sound travels several feet through air before reaching your ears. Your head and outer ears (pinnae) shape the sound in specific ways. You get reflections off walls, floors, ceilings. Your whole body feels bass frequencies. It's a complete acoustic experience.

With headphones, the driver is maybe an inch from your eardrum. There's no room, no natural acoustic space. The sound doesn't interact with your outer ear the same way. You don't feel bass with your chest. You only hear it.

This creates some interesting challenges:

Bass needs artificial boost. In a room with speakers, bass frequencies get reinforced naturally. With headphones, if you tune them "flat," they sound thin and bass-light. We'll talk about target curves `(Chapter 2)` later, but just know: headphones that measure flat don't sound neutral.

Proximity matters. Because the driver is so close to your ear, tiny changes in distance create big changes in sound. This is why ear pad thickness `(Chapter 10)` is a real tuning parameter, not just a comfort choice.

The sealed environment changes everything. Most headphones create at least a partial seal around your ear. This affects bass response dramatically. We're not working with open air like speakers. We're working with an acoustic chamber.

The Complete System

Here's something that took me years to really internalize: you're building a whole system, not just picking a driver.

Think of it this way: **Driver + Ear Cup + Ear Pads + Your Ear Canal = The Sound You Hear**

Change any one of these elements, and you change the entire sound. This is why you can't just grab a good driver, stick it in any enclosure, and expect magic. It all has to work together.

The ear cup isn't just a housing. It's an acoustic chamber (Chapter 9). Its volume, shape, and internal surfaces all affect the sound. Too small and your bass rolls off early. Too large and you might get resonances that make things sound weird.

The ear pads determine how far the driver sits from your ear. They also affect the seal. A better seal means better bass. Different pad materials absorb different frequencies differently. Even the depth of the pad matters. We'll cover all of this in Ear Pads and Comfort (Chapter 10).

Damping materials inside the cup absorb certain frequencies, tame resonances, and shape the overall response. This is one of your most powerful tuning tools, and one of the easiest to get wrong. More on this in Damping Materials (Chapter 11).

What This Means When You're Building

Understanding the system nature of headphone design changes how you approach building. Instead of "which driver sounds best?" the question becomes "which driver works best in this system?"

A driver that sounds amazing in one cup design might sound mediocre in another. The same driver with different pads can sound dramatically different. I've seen $15 budget drivers outperform $150 premium drivers when the budget driver happened to match the enclosure better.

This is actually good news for DIY builders. That $15 driver winning wasn't luck. It fit the cup. Matching beats spending.

A Few Things I Wish Someone Had Told Me

There's no "correct" sound. The headphone industry has spent decades arguing about what headphones should sound like. The truth is, it's complicated. What sounds great to one person sounds wrong to another. We have better targets now than we did 20 years ago (the Harman curve (Chapter 2) research is worth knowing), but ultimately you're building headphones that sound good to you.

Measurements and listening both matter. Some people in the DIY community worship measurements. Others ignore them entirely and just listen. The reality is you need both. Measurements tell you what's happening objectively. Listening tells you whether it matters to your ears. Neither alone is complete.

Your first build won't be perfect, and that's fine. I've built headphones for 25+ years and I still have to iterate. Your first pair is a learning experience. Plan for it, embrace it, and don't beat yourself up when things don't sound exactly right on the first try. That's how this works.

What's Next

So how do we describe and measure what we're hearing? That's where frequency response comes in.

In Understanding Frequency Response (Chapter 2), we'll explore what those graphs actually show, why different headphones sound different, and why a "flat" frequency response isn't what you think it is.

Or, if you're itching to get hands-on, you could also jump straight to Your First Build and come back to the theory later. There's no wrong path here.

CHAPTER 2

Understanding Frequency Response

Learn how to read frequency response graphs, why "flat" isn't good for headphones, and what the Harman target curve means for DIY builders.

9 min read   makerphones.com/learn/understanding-frequency-response

If you've spent any time looking at headphones online, you've probably seen frequency response graphs. They look intimidating, these squiggly lines with numbers and colors. But once you can read one, it's one of the most practical skills you've got as a headphone builder. Here's how to read these graphs and, more importantly, what they actually tell you about sound.

What Is Frequency Response?

Frequency response is a measurement of how loud a headphone plays different frequencies. That's it. It's not magic, it's not subjective, it's just "how much sound pressure does this headphone produce at each frequency?"

When you look at a frequency response graph, the horizontal axis shows frequency (from deep bass around 20Hz up to the highest treble at 20kHz). The vertical axis shows the sound pressure level in decibels (dB). Higher on the graph means louder at that frequency.

So when the line goes up, that means more of that frequency. When it dips down, there's less of that frequency. The shape of this line tells you a lot about how a headphone will sound.

The Three Main Regions

Let's break the frequency spectrum into three main areas.

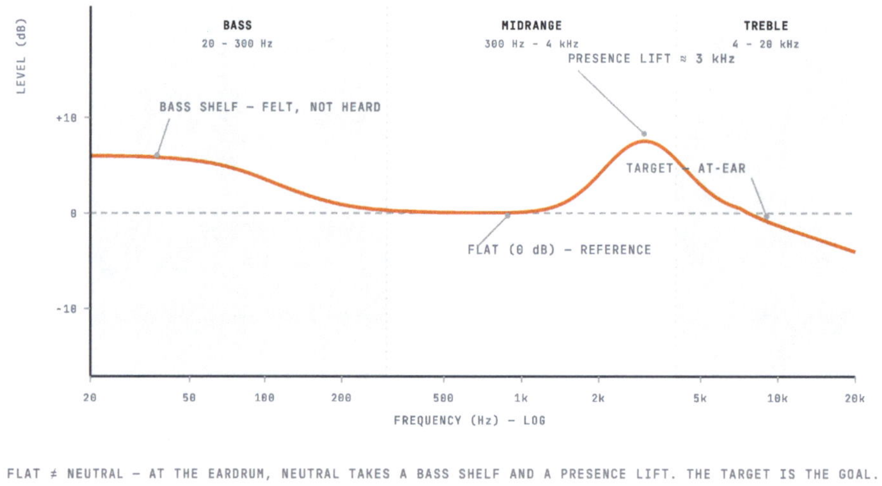

Fig. 2.1 – Reading a frequency-response curve: the three regions, a flat reference line, and an at-ear target (illustrative) showing why flat isn't neutral.

Bass (20Hz - 300Hz)🔗

Bass is the thump, the rumble, the low end you feel. The very deepest bass (20–60Hz) is sub-bass, you feel this more than hear it in speakers, but with headphones it's purely audible. Mid-bass (60–250Hz) is where kick drums, bass guitars, and the body of most bass instruments live.

Here's something critical: with headphones, you generally want the bass elevated compared to the midrange. If a headphone measures "flat" in the bass, it's going to sound thin and anemic. Remember from How Headphones Create Sound (Chapter 1), we don't have room gain or physical impact with headphones, so we need to compensate.

Most good headphones have a gentle bass shelf: meaning the entire bass region is lifted by a few dB compared to the mids. This sounds natural to our ears, even though it's technically not "flat."

Midrange (300Hz - 4kHz)🔗

Most of the music lives here. Vocals, guitars, pianos, most instruments, they all have their fundamental frequencies in the midrange. This is the most critical region to get right because our ears are incredibly sensitive to midrange problems.

Within the midrange, there are some sensitive spots:

Lower mids (300Hz – 1kHz): Warmth and body. Too much here and things sound muddy or boomy. Too little and music sounds thin.

Upper mids (1kHz – 4kHz): Presence and clarity. This region is critical for speech intelligibility and the "forward" or "in your face" character of headphones. A peak around 3kHz sounds harsh and fatiguing. A dip sounds recessed or laid-back.

Treble (4kHz - 20kHz)

This is where things get complicated. The treble region determines how "airy," "detailed," or "sparkly" headphones sound, but it's also where things go wrong most often.

The treble interacts heavily with your ear anatomy. Specifically, the shape of your outer ear (the pinna) creates peaks and dips in the treble as you move your head. This is what helps us localize sound in the real world, but it makes headphone measurement tricky.

Generally, you want the treble to follow a gradual downward slope above 10kHz. Sharp peaks anywhere in the treble sound harsh and fatiguing. A big dip around 4–8kHz sounds dark and muffled.

Why "Flat" Isn't Good for Headphones

This is probably the most important concept in headphone acoustics. A "flat" frequency response, where all frequencies are reproduced at exactly the same level, sounds wrong on headphones.

Why? Because that's not how we hear in the real world.

When we listen to music through speakers in a room, several things happen that headphones don't replicate:

- Bass frequencies are reinforced by room resonances (adds 3–6dB of bass in most rooms)
- High frequencies above 10kHz roll off naturally with distance
- The sound interacts with our outer ears, adding pinna-related coloration
- We perceive sound from outside our head, not inside it

Headphones that measure flat sound bass-light, midrange-heavy, and overly bright. To sound "neutral", meaning similar to good speakers in a good room, headphones need a specific non-flat response.

The Harman Target Curve

In the early 2010s, audio researchers at Harman International (JBL, AKG, etc.) did something invaluable: they ran extensive studies to find out what frequency response people actually prefer in headphones. Thousands of listeners, hundreds of headphones, statistically rigorous methodology.

The result is the Harman target curve: a specific frequency response shape that most people prefer when they don't know what they're listening to. It's not what everyone prefers (bass heads want more bass, some people prefer brighter signatures), but it's a well-validated starting point. The research behind it, and where to read it, is mapped in Sources & Further Reading.

Key characteristics of the Harman target:

- Bass elevated by about 5–6dB below 200Hz
- Gradually rising response through the mids
- Peak around 3–4kHz (presence region)
- Gradual roll-off above 10kHz

For DIY builders, that's a gift. Instead of guessing what "sounds right," you have a target to aim for. When you're tuning your damping materials `(Chapter 11)`, you can measure your headphones and compare to the Harman target to guide your decisions.

How to Read FR Graphs

Here's a practical guide to interpreting graphs.

Look at the overall shape before you chase any single wiggle. Is the bass elevated? Does it roll off gradually going toward the mids? Is there a presence peak? Does the treble slope downward? These big-picture characteristics matter more than small wiggles.

Peaks are problem areas. Any sharp spike in the frequency response, especially narrow spikes that go up 5dB or more above the trend, will be audible. Peaks in the 2–8kHz range are especially noticeable. In DIY headphones, peaks often indicate resonances in the acoustic chamber or the driver itself.

Dips are less audible than peaks. Our ears are less sensitive to missing frequencies than to excess frequencies. A 6dB dip is less objectionable than a 6dB peak, all else equal. That said, big dips in the mids (the "hollow" sound) are very noticeable.

Channel matching matters. When you see two lines (one for each ear), they should be very close together, ideally within 2–3dB across most of the frequency range. Large differences between channels mess with imaging and stereo perception.

Smoothing hides details. Frequency response graphs are often smoothed to make them easier to read. More smoothing makes things look cleaner but hides narrow resonances. When you're tuning your own builds, look at both smoothed and unsmoothed views.

Practical Application for Builders

Here's how this knowledge applies when you're actually building:

1. Don't panic about small deviations. Getting within 5–6dB of your target curve over most of the frequency range is a win. Chasing perfect measurement is an endless road.

2. Fix the big problems first. If you have a 15dB peak at 8kHz, that's the problem to solve. Don't worry about the gentle bass rolloff until the obvious problems are fixed.

3. Measurements guide your listening. After any damping change, measure AND listen. Sometimes a measurement improvement doesn't sound better subjectively. Trust your ears too.

4. Focus on problem areas first. Don't try to match a target curve exactly from the start. Fix the glaring problems, harsh peaks, boomy bass, recessed mids, then fine-tune.

5. Listen more than you measure. Seriously. I've spent days chasing measurements only to realize I preferred an earlier version that measured "worse." Trust your ears.

What's Next

The next specs to get straight are the ones that decide whether a driver will even play loud enough on your gear. In Impedance and Sensitivity (Chapter 3), we'll demystify these specs and show you why they matter more than most people think.

Or, if you want to get hands-on with measurement right away, jump to Budget Measurement Setup (Chapter 19) to learn about affordable tools that will let you see what your headphones are actually doing.

Impedance and Sensitivity

Learn what impedance and sensitivity actually mean for headphone builders, why these specs matter, and how to match drivers to your amplifier.

7 min read makerphones.com/learn/impedance-and-sensitivity

When you're looking at driver specs, two numbers always show up: impedance and sensitivity. Most people gloss over these or assume "higher is better" (or lower, depending on what they've heard). Get these wrong and your headphones either won't get loud enough or won't play nice with your amp. So let's figure out what the numbers actually mean.

What Impedance Really Measures

Impedance is measured in ohms (Ω), and you'll see headphones ranging from 8Ω up to 600Ω. The common wisdom is "low impedance is easier to drive, high impedance needs a powerful amp." That's partially true, but it's not the complete picture.

What Impedance Actually Is

Technically, impedance is the opposition to alternating current (AC) flow in a circuit. It's like resistance, but more complex because it varies with frequency. In a dynamic driver, impedance comes from three main sources:

The voice coil resistance: This is just the DC resistance of the copper wire in the coil. It's constant across all frequencies.

Inductance: The coil is, well, a coil, which means it has inductance. This causes impedance to increase at high frequencies.

Mechanical impedance: The moving parts of the driver (diaphragm, suspension, voice coil mass) also present impedance that varies with frequency. This creates the impedance peak you see near the resonant frequency.

Why Impedance Varies with Frequency

When you see a headphone listed as "150Ω," that's the nominal impedance: a single number representing a complicated curve. If you measure impedance across the full frequency range, you'll see it's NOT constant.

There's a big peak at the driver's resonant frequency (Fs), often 50–150Hz for headphone drivers. Below that, impedance drops. Above that, impedance gradually rises due to inductance. The "nominal" impedance is typically measured at 1kHz, where things are relatively stable.

This matters because your amplifier's output impedance interacts with the headphone impedance. If your amp has high output impedance (common in tube amps), the frequency response actually changes based on the impedance curve. A 300Ω headphone on a high-impedance amp will sound different than the same headphone on a low-impedance amp.

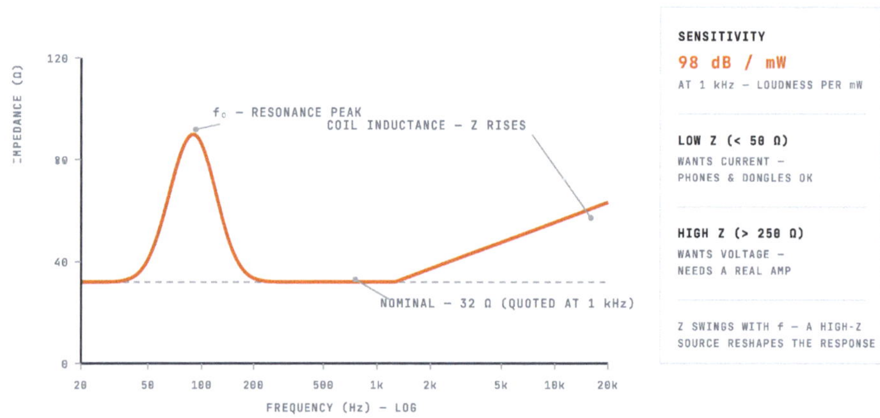

IMPEDANCE IS A CURVE, NOT A NUMBER – THE PEAK AND THE HF RISE ARE WHY SOURCES CHANGE THE SOUND.

Fig. 3.1 – Impedance is a curve, not a number (illustrative): the resonance peak, the inductive rise, and what sensitivity and impedance mean for amp matching.

Practical Implications for DIY Builders

Portable use (phone, laptop): Stick to 32Ω or lower. These sources have high output impedance and limited current. High-impedance headphones won't be driven properly.

Desktop use with dedicated amp: Any impedance works. Higher impedance (150–300Ω) can have advantages: more turns of wire in the voice coil, potentially better high-frequency response.

First build: Use 32Ω. It's the universal value that works everywhere. You can experiment with impedance later once you understand the system better.

Sensitivity: How Loud for How Much Power

Sensitivity tells you how much sound pressure a driver produces for a given electrical input. It's usually expressed as dB SPL per milliwatt (dB/mW) or dB SPL per volt (dB/V) at 1kHz.

You'll see values like "103 dB/mW" or "110 dB/V." Higher number = louder for the same input. But wait, dB/mW and dB/V are different scales, and you can't compare them directly without knowing the impedance. This trips up a lot of people.

Making Sense of the Numbers

Here's what matters practically:

For portable use (phone/laptop): You need high sensitivity, around 100 dB/mW or higher. These sources are power-limited. Low sensitivity means low volume.

For desktop use with amplification: Sensitivity matters less. Almost any driver will get loud enough with a proper amp.

Matching to your source: If you're building for a specific source (like using a portable DAC/amp), check its maximum output power and calculate whether your driver will reach comfortable listening levels.

The Sensitivity-Impedance Interaction

Here's where it gets interesting. Sensitivity and impedance are related through the driver's efficiency. A driver with the same efficiency but different impedance will measure differently in dB/mW vs. dB/V.

Without getting deep into the math: if you're comparing drivers for portable use, use dB/mW. If you're comparing for desktop use with voltage-output amps, use dB/V. If you only have one measurement available, note which standard it uses.

Matching Drivers to Amplifiers

All that theory collapses into a short decision tree:

Building for phone/laptop use:

- Impedance: 16–32Ω
- Sensitivity: 100 dB/mW or higher
- Result: Loud enough, no dedicated amp needed

Building for portable DAC/amp (FiiO, DragonFly, etc.):

- Impedance: 32–150Ω works fine
- Sensitivity: 95 dB/mW or higher
- Result: Good match for most portable gear

Building for desktop setup with dedicated amp:

- Impedance: Anything from 32–600Ω
- Sensitivity: Less critical, amp compensates
- Result: Maximum design flexibility

Common Misconceptions

⚠ CAUTION

"Higher impedance sounds better." No. Impedance affects system matching, not inherent sound quality. A well-designed 32Ω driver can sound better than a poorly-designed 300Ω driver.

"Sensitivity is fixed." Nope. It varies with frequency. The listed sensitivity is nominal, a representative value. The actual sensitivity curve can vary significantly, especially around the driver's resonant frequency.

Measuring Impedance and Sensitivity

If you want to verify your driver specs (or measure a salvaged driver with unknown specs), you can do this relatively easily.

For impedance: Use a multimeter to measure DC resistance. This gives you the resistive component, which is close to the nominal impedance for most drivers.

For sensitivity: You need a calibrated microphone, a test rig, and measurement software (like REW (Chapter 19)). Play a sine wave at 1kHz, measure the output SPL, note the input voltage or power, and calculate sensitivity.

What's Next

Next up: open-back vs closed-back. It's the first big fork in any build, and it changes how everything else gets tuned. Open vs Closed Back Design (Chapter 4) walks through how the choice ripples out into sound quality and tuning complexity.

Or jump to the Driver Selection Guide (Chapter 7), where these specs turn into an actual buying decision.

Open vs Closed Back Design

Learn the real differences between open-back and closed-back headphones: how they sound, how to design them, and which is right for your first build.

6 min read   makerphones.com/learn/open-vs-closed-back-design

One of the first big decisions you'll make when designing headphones is whether to go open-back or closed-back. It's not an aesthetic call or a minor tweak. It changes how the headphone sounds, how hard it is to tune, and what situations it's good for. Here are the real differences, and how to decide which approach makes sense for your build.

The Basic Difference

Closed-back headphones have a sealed rear cup. The back of the driver is enclosed in an airtight (or nearly airtight) chamber. Sound can't escape out the back, and outside sound can't get in.

Open-back headphones have vents, grilles, or completely open rear cups. The back of the driver is exposed to the outside air, either directly or through acoustic openings.

But the acoustic implications of this choice are profound.

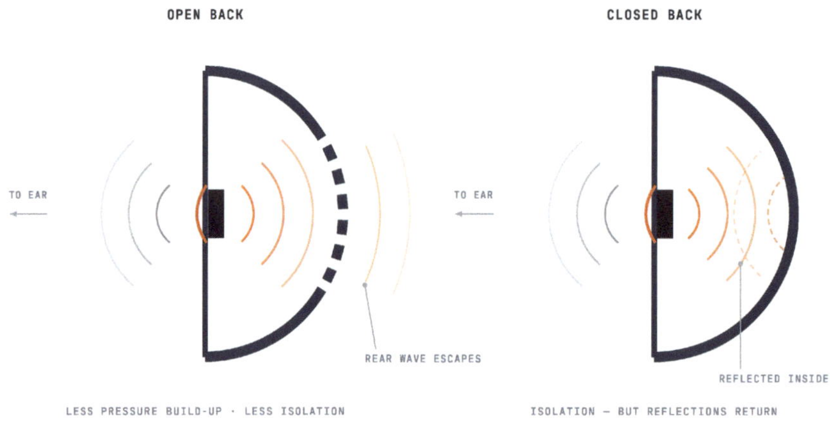

Fig. 4.1 — The same driver, two rear-cup decisions, and where the rear wave goes.

How Open-Back Works

When a driver moves, it creates sound waves on both sides of the diaphragm. The front wave (toward your ear) is what you're trying to hear. The back wave (behind the driver) is moving in the opposite direction.

In an open-back design, that back wave isn't trapped. It radiates out into the room. This creates several effects:

No pressure buildup: Because the back is open, there's no chamber for pressure to build up in. This generally makes bass response more linear and predictable, but it also means you lose the bass reinforcement that a sealed chamber provides.

Less resonance: Without an enclosed chamber behind the driver, you don't get standing waves and chamber resonances that need to be controlled.

Natural, spacious sound: Open-backs tend to have a more "speaker-like" or "airy" sound. The soundstage often feels wider and less in-your-head. Many audiophiles strongly prefer this character for critical listening.

Sound leaks both ways: People near you can hear what you're listening to. You can hear what's happening around you. Not good for commuting or open offices.

How Closed-Back Works

In a closed-back design, the back wave is trapped in the sealed chamber. This creates a different set of acoustic phenomena:

Back pressure matters: The sealed air behind the driver acts like a spring, affecting the driver's resonant frequency and compliance. Chamber volume directly affects bass extension: larger chamber generally means deeper bass.

Resonances are a challenge: The trapped air in the chamber creates standing waves at specific frequencies. These appear as peaks in the frequency response that need to be controlled with damping materials `(Chapter 11)`.

Isolation: A well-sealed closed-back provides 10–25dB of passive noise isolation. Good for noisy environments, bad for spatial awareness.

Bass can be stronger but less controlled: The sealed chamber can reinforce bass, but it can also create boomy, one-note bass if not tuned properly. Getting clean, extended bass in a closed-back is harder than it sounds.

Sound Character Differences

These aren't universal rules, good design can overcome typical limitations of each type, but here's what you'll generally find:

Open-back tends toward: Wider soundstage, more natural timbre, better imaging, less fatiguing for long listening sessions, more speaker-like character.

Closed-back tends toward: More bass impact, more intimate sound, better isolation from environment, more "in your head" presentation.

The very best headphones in the world are open-backs: the Sennheiser HD800, Audeze LCD-4, HiFiMAN Susvara. But that doesn't make closed-back inferior for all uses.

DIY Difficulty Comparison

Here's something that might surprise you: **open-back is generally easier to get right on your first build.**

Why open-back is easier:

- You don't have to control rear-chamber resonances

- Bass response is more predictable and linear
- Less sensitive to cup volume, moderate changes don't dramatically shift bass
- Easier to iterate, more forgiving of design variations
- Your biggest challenge is driver selection and pad choice, both manageable

Why closed-back is harder:

- Chamber resonances need to be actively controlled with damping
- Bass tuning is more complex: chamber volume, porting, and damping all interact
- Getting a proper seal for good bass while maintaining comfort is tricky
- More variables to manage simultaneously

This is why I recommend starting with open-back for your first design. You'll learn more with less frustration.

Semi-Open: The Middle Ground

Some designs fall between fully open and fully closed. These use vents or ports to control how much the back wave escapes.

Semi-open designs can offer some of the soundstage character of open-backs while providing more bass than a fully open design. However, they're harder to predict and tune. Small changes to the vent size or placement have big effects.

I'd recommend understanding both open and closed designs before experimenting with semi-open. It adds complexity without a clear "right answer."

Choosing for Your First Build

Here's a practical decision guide:

Choose open-back if:

- This is your first DIY headphone
- You'll primarily listen at home
- You value soundstage and natural sound over isolation
- You want the easier design challenge

Choose closed-back if:

- You need noise isolation (office, travel, recording studio)
- You specifically want more bass impact
- You're okay with more tuning complexity

- You've already built a successful open-back

What's Next

Open vs. closed is just one dimension of the design space. The next one is form factor: over-ear, on-ear, or in-ear.

In Headphone Form Factors `(Chapter 5)`, we'll cover over-ear, on-ear, and in-ear designs, and explain why over-ear is almost always the right choice for DIY builders starting out.

When you're ready to start thinking about actual components, head to Your First Build for a beginner-friendly open-back project.

CHAPTER 5

Headphone Form Factors

Over-ear, on-ear, or IEM: not all headphone form factors are equally suited for DIY building. Here's what you need to know before starting your first project.

5 min read   makerphones.com/learn/headphone-form-factors

Walk into any audio store and you'll see headphones in all shapes and sizes: massive over-ears, compact on-ears, tiny earbuds. To a listener that's a comfort or portability call. To a builder, the form factor drives everything downstream: acoustic design, driver selection, comfort engineering, and build complexity. Here's what each of the main categories means for DIY projects.

The Three Main Categories

Headphones fall into three broad form factors:

Over-ear (circumaural): The ear cups completely surround your ears, resting on your head around the ear rather than on the ear itself.

On-ear (supra-aural): The ear cups rest directly on your ears, pressing against them rather than surrounding them.

In-ear (IEMs, In-Ear Monitors): Sit inside your ear canal, creating a seal with tips or custom molds.

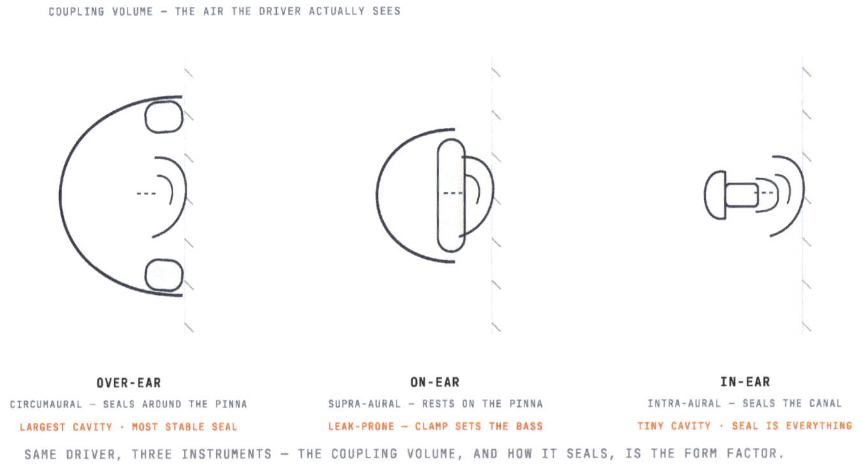

Fig. 5.1 – Over-ear, on-ear, and in-ear against the same ear: what changes is the coupling volume, the air the driver actually sees, and how it seals.

For DIY building, they're not equally accessible.

Over-Ear: The DIY Sweet Spot

Most DIY headphone building happens here, and the reasons stack up fast.

Generous size: You have physical space to work with. Mounting a 40–50mm driver is straightforward. You can fit damping materials, experiment with chamber volumes, and make adjustments without working at microscopic scale.

Forgiving fit: Because the cups surround the ear rather than pressing on it, small variations in shape or size are more tolerable. Your design doesn't need to be as precisely fitted to each individual person.

Design flexibility: You can go open or closed, try different pad materials, experiment with damping: all the main tuning tools are available to you.

Standard components: 40mm and 50mm drivers have a huge selection. Pads are standardized enough that you have many options. Headband designs are well-documented.

Most of the content on this site focuses on over-ear designs. If this is your first build, over-ear is the right choice.

On-Ear: More Challenging

On-ear headphones look similar to over-ear from a distance, but the acoustics are quite different.

The seal problem: Because the pad rests on the ear rather than around it, getting a consistent seal is much harder. The bass response depends heavily on how well and consistently the pad seals against each user's ear. This varies significantly person to person.

Comfort engineering is critical: On-ear headphones press directly on your ear. Getting the clamp force and pad pressure right matters enormously. Too much pressure is painful for long sessions. Too little and the seal (and bass) suffers.

Smaller drivers: On-ear cups are smaller, which means smaller drivers, typically 30–40mm. Less driver area generally means less bass capability.

On-ear is workable for DIY, but it adds complexity without a clear performance advantage. I'd suggest building an over-ear first and exploring on-ear later.

In-Ear Monitors: A Completely Different World

IEMs are phenomenal audio devices, but they're a completely different design challenge from headphones. Building your own IEMs is an advanced project with a very different skill set.

Tiny drivers: IEM drivers are typically 6–10mm dynamic drivers, or balanced armature drivers that are even smaller. Sourcing, handling, and mounting these requires precision that's significantly more demanding.

Shell design: The IEM body needs to fit precisely in or around your ear canal. Custom IEMs require ear impressions and careful fitting. Universal IEMs need to work for a wide range of ear canal sizes.

Crossovers: Multi-driver IEMs (which is where the really interesting designs live) require acoustic crossovers to divide frequencies between drivers. This is its own area of expertise.

I love IEMs and I plan to cover them in future content. But they're not a starting point. Master over-ear first, then consider IEMs as a separate advanced project.

Driver Size within Over-Ear

Within over-ear headphones, there's another size decision: 40mm vs. 50mm drivers.

40mm drivers:

- Most common, largest selection of options
- Lighter, smaller cups possible
- Slightly less bass extension potential (smaller diaphragm area)
- Easier to source at reasonable prices

50mm drivers:

- Larger diaphragm = more bass potential
- Requires larger cups (heavier)
- Fewer options at budget price points
- Preferred by some premium manufacturers for bass impact

For a first build: 40mm. The selection is better, the cups can be lighter, and the bass difference is smaller than marketing would suggest.

Common Mistakes

⚠ DESIGNING THE CUP BEFORE MEASURING THE ACTUAL DRIVER

Not all `40mm` drivers are exactly 40mm. Flange diameter, mounting depth, and overall dimensions vary between manufacturers. Order your drivers first, measure the real parts, then design your cup around them, not around the nominal spec.

⚠ CHOOSING A 50MM DRIVER ASSUMING MORE BASS

The reasoning sounds right, bigger driver, more bass. But in a DIY build, bass depends far more on enclosure design and tuning than on driver diameter. The extra bass from `50mm` is real but smaller than you'd expect, and on a first build it's not worth the heavier cups and thinner sourcing options.

What's Next

Form factor sets the stage; the driver is what actually makes sound. In Driver Technologies (Chapter 6), we'll cover dynamic, planar magnetic, and other driver types, and explain why dynamic drivers are the right choice for most DIY builders.

CHAPTER 6

Driver Technologies

Learn about dynamic, planar magnetic, balanced armature, and electrostatic drivers, and why dynamic drivers are the right choice for DIY builders.

7 min read   makerphones.com/learn/driver-technologies

If you've been following along from the beginning, you already know the basics of how dynamic drivers work. But there are actually several different driver technologies used in headphones, each with its own physics, sound characteristics, and practical considerations for builders. There are four you'll run into, and only one is really worth your time as a builder.

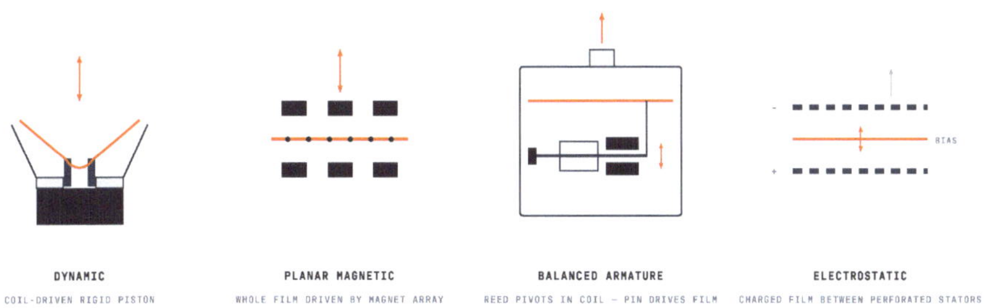

Fig. 6.1 – Four ways to move air: the moving element in orange.

Dynamic Drivers: The DIY Standard

We covered the basics in How Headphones Create Sound (Chapter 1), but let's go deeper into why dynamic drivers dominate both commercial headphones and DIY builds.

A dynamic driver uses a voice coil suspended in a magnetic gap, attached to a diaphragm. Current through the coil creates a magnetic field that interacts with the permanent magnet, moving the diaphragm and creating sound.

Key components in detail:

The magnet structure: Usually neodymium (for strength and low weight) or ferrite (cheaper, heavier). Modern designs use carefully shaped pole pieces to concentrate the magnetic field precisely where needed.

The voice coil: Wound from copper or aluminum wire. Copper sounds slightly warmer, aluminum is lighter (better high-frequency response). The coil diameter, length, and wire gauge all affect performance.

The diaphragm: Common materials include paper/cellulose (warm, natural damping), Mylar/PET film (clear, neutral, very common), and various composites. Each has tradeoffs between rigidity, weight, and internal damping.

The suspension: The surround (outer edge) and spider (inner suspension) keep the voice coil centered while allowing movement. Compliance, how easily the driver moves, affects bass extension and sensitivity.

Why dynamic drivers win for DIY:

- Available in huge variety from many suppliers
- Affordable: good drivers from $5 to $50
- Self-contained: no external electronics needed
- Well-understood behavior
- Works with any standard amplifier
- Replaceable if damaged

Planar Magnetic Drivers: Interesting But Challenging

A planar magnetic driver ditches the voice coil. Instead, a planar driver has a thin membrane with conductor traces (like circuit board traces) running across it. This membrane sits between two sets of magnets.

When current flows through the conductor traces, the entire membrane moves at once, where a dynamic driver only drives the center. Proponents argue that driving the whole surface simultaneously reduces distortion and improves transient response.

The appeal: Planar magnetic headphones (Audeze, HiFiMAN, Meze) are known for their low distortion, good transient response, and "effortless" sound character. Some of the world's most respected headphones are planar magnetic.

The DIY challenge:

- Extremely difficult to source as raw drivers
- Require powerful amplification (low sensitivity, complex impedance)
- Large and heavy
- Very expensive when available
- Not practical for first builds

DIY planar builds exist, but they're advanced projects requiring significant resources and expertise. Don't start here.

Balanced Armature Drivers: IEM Territory

Balanced armature (BA) drivers are the technology that makes modern multi-driver IEMs possible. They're tiny, sometimes just a few millimeters, and incredibly efficient. Almost all hearing aids use balanced armature technology.

A BA driver uses a small armature (a metal bar) balanced between two magnets. When current flows through a coil around the armature, it tilts toward one magnet or the other, driving a small diaphragm through a mechanical linkage.

BA advantages:

- Extremely compact
- Very efficient (high sensitivity)
- Multiple drivers can be combined for full-range coverage

BA limitations for DIY:

- Extremely small, requiring precision work
- Limited bass extension (most need dynamic drivers for bass)
- Require crossovers for multi-driver systems
- Sourcing is difficult and expensive

For IEM building, BA drivers are essential knowledge. For over-ear headphones, they're irrelevant.

Electrostatic Drivers: Audiophile Stratosphere

Electrostatic headphones throw out magnets entirely and run on static charge. An extremely thin membrane (just microns thick) is suspended between two perforated metal plates called stators. The membrane is given a static electrical charge (hence the name).

When audio signal is applied to the stators, it creates an alternating electrical field. The charged membrane is attracted to one stator and repelled by the other, moving back and forth to create sound.

Why electrostats are special: The membrane is so light (often just a few micrograms) that it can respond to extremely fast transients. Many audiophiles consider top electrostatic headphones (Stax SR-009, HiFiMAN Shangri-La) the best-sounding headphones ever made, full stop, and they're not entirely wrong.

Why they're not DIY-accessible:

- Require dedicated high-voltage amplifiers ("energizers"), not compatible with standard amps
- Membrane fabrication is extremely specialized
- High voltages involved are potentially dangerous
- No practical component sourcing for DIY builds
- Extremely expensive even for commercial products

Electrostatics are fascinating technology worth understanding, but building one yourself is a serious engineering project well past the beginner or intermediate range.

So Which One Do You Build With

For your first build, and probably your next several, use a dynamic driver. Everything else here is worth knowing, but none of it belongs in a beginner's enclosure. Dynamic drivers are cheap enough to experiment with, work with any amp, and have an active DIY community that's already solved most of the problems you'll hit.

The other technologies are interesting and worth learning about, but they're not practical starting points. Master dynamic drivers first. The principles you learn, acoustic design, damping, tuning, transfer to other technologies if you eventually explore them.

Common Misconceptions

> ⚠ CAUTION
>
> **"Bigger drivers sound better."** Not necessarily. Larger drivers can move more air (good for bass) but can have breakup modes (bad for highs). It's about implementation.
>
> **"Expensive drivers guarantee good sound."** Nope. System integration matters more than driver cost. I've heard $200 drivers sound terrible in poor enclosures and $15 drivers sound great in well-designed ones.
>
> **"Measurements tell you everything."** They tell you a lot, but not everything. Transient response, distortion characteristics, and how the sound interacts with your specific ear anatomy all matter.

What's Next

Knowing dynamic is the right call still leaves you staring at hundreds of them. In Driver Selection Guide `(Chapter 7)`, we'll walk through how to actually choose one for your specific project: size, impedance, material, supplier, and budget all considered.

PART

2

Components & Materials

Driver Selection Guide

Learn how to choose the right drivers for your DIY headphone build: size, impedance, materials, and specific recommendations for beginners.

7 min read makerphones.com/learn/driver-selection-guide

Standing in front of Parts Express's website or scrolling through AliExpress listings, you hit the same wall everyone does: there are hundreds of options. How do you choose? What actually matters? Here's the decision process I use when selecting drivers, based on 25+ years of working with headphones.

Start with Your Use Case

Before you even look at specs, answer these questions:

Where will you use these headphones? Home only, or portable use? This determines impedance (Chapter 3) requirements. Home use with a dedicated amp? You can go up to 300Ω. Portable use with phones/laptops? Stick to 32Ω.

What's your sound signature preference? Neutral and accurate, or fun and colored? Warm and smooth, or bright and detailed? This guides driver material and design choices.

What's your budget? Be realistic. You can build excellent headphones with $15 Dayton Audio drivers, or spend $100+ on premium drivers. Neither is "wrong." They're different starting points.

Is this your first build? If yes, choose forgiving drivers. Well-reviewed, popular options from known manufacturers. Save the exotic experiments for build #3.

Open-back or closed-back? As we discussed in Open vs Closed Back Design (Chapter 4), this affects everything.

The Key Specs to Evaluate

Driver Diameter: 40mm vs. 50mm

As we covered in Headphone Form Factors (Chapter 5), 40mm is the right choice for most first builds. Larger selection, smaller cups possible, easier to work with. The bass difference is real but smaller than marketing claims.

Impedance: Match to Your Source

For portable use: 32Ω. Full stop. For desktop use with amplification: 32–150Ω works well for most setups. Higher impedance (250–300Ω) can be excellent but requires more amp power.

Sensitivity: How Loud Will They Get?

For portable use, you need at least 95 dB/mW, ideally 100+ dB/mW. For desktop use with a proper amp, sensitivity matters less. Your amp compensates.

Frequency Response: What Does It Show?

Look for measurements if available (Parts Express usually provides them). You want: smooth response without major peaks or dips, reasonable bass extension (usable down to 30–40Hz), no obvious treble problems.

Remember, you'll be tuning the final response with your enclosure and damping. The driver's raw response is just the starting point. No need to chase a perfectly flat raw curve.

THD (Total Harmonic Distortion)

Lower is generally better, but for headphone drivers, distortion at normal listening levels is usually not the limiting factor in sound quality. Don't obsess over this spec if other factors check out.

Recommended Drivers for First Builds

Budget Tier ($10-20 per pair)

Dayton Audio CE38MB-32 (38mm, 32Ω): The one part in Dayton's catalog actually marketed as a headphone replacement driver, and remarkably good for the price. A solid first driver: decent frequency response, forgiving of enclosure variations, and cheap enough to buy extras for experimentation. One build note: the Mylar diaphragm is fragile, and Dayton specifies installing it behind a grille. Design your baffle accordingly.

A word about the rest of Dayton's CE series, because the names look tempting: it's mostly small *loudspeaker* drivers, 4 and 8 ohm, paper or aluminum cones, that happen to be headphone-sized. The tell for headphone use is 32-ohm impedance with a Mylar diaphragm, the "MB-32" type like the CE38MB-32. Don't drop a 4 or 8 ohm cone driver into a headphone build and expect headphone behavior.

Mid-Range Tier ($20–50 per pair)

Salvaged drivers from quality commercial headphones: Surprisingly good option. A Sennheiser, Audio-Technica, or AKG headphone with broken headband/cable often has a perfectly good driver. Thrift stores and eBay are your friends. The downside: you don't get spec sheets.

Premium Tier ($50–100+ per pair)

At this level, you're buying drivers used in mid-range commercial headphones. You'll find some options on AliExpress from Chinese suppliers, but quality control is inconsistent. Research thoroughly before spending significant money.

Honestly? For your first build, don't spend more than $30 on drivers. The enclosure design and tuning matters more. Save the premium drivers for a build where you have the skills to let them shine.

AliExpress: Proceed with Caution

AliExpress has inexpensive headphone drivers, and some are genuinely good. But the quality control is wildly inconsistent. The driver you buy may or may not match the advertised specs. Channel matching between left and right can be poor. Fakes of popular driver models exist.

If you use AliExpress:

- Buy from sellers with thousands of positive reviews
- Check if measurements are available online from other builders
- Order extras, expect some duds
- Never use them for a build you care about until you've verified the specific lot

For your first build, stick with Parts Express, Madisound, or other established audio component suppliers. The extra cost is worth the reliability.

Evaluating Salvaged Drivers

If you're using drivers salvaged from commercial headphones:

Verify they work: Connect to an amp and play audio before you spend time designing around them.

Check channel matching: Do both sound similar? A multimeter can check DC resistance. They should be within 1–2Ω of each other for a matched pair.

Measure if possible: If you have measurement equipment, take baseline measurements before building. This helps you understand what you're working with.

Document the source: Write down where you got them. If you ever want to source more, you'll need to know.

What to Actually Do

Don't experiment with unknowns: Your first build should use a driver with reviews or measurements from other builders. Head-Fi, diyAudio forums, and Reddit r/headphones all have discussions of specific drivers.

Stay mainstream on impedance: 32Ω. Don't go exotic.

Start with 40mm: Lighter, cheaper, easier to work with than 50mm.

Buy from known suppliers: Parts Express, not AliExpress. Pay a bit more for reliability.

Don't overspend: Your first build is for learning. Save the $100 drivers for build #3 when you know what you're doing.

Consider buying extras: If drivers are cheap enough, buy 2 pairs. Having spares lets you experiment without fear.

What's Next

Picking the driver is only half the shopping trip. You still need cups, pads, foam, wire, and a screw count, and the difference between one cart and three is knowing which supplier carries what. In Sourcing Components (Chapter 8), we'll cover suppliers, what to buy, and how to build a complete parts list for your project.

CHAPTER 8

Sourcing Components

A complete guide to sourcing everything you need for DIY headphones: drivers, pads, cables, materials, and hardware. Includes specific suppliers and recommendations.

7 min read   makerphones.com/learn/sourcing-components

You know what drivers you want. You have a design in mind. Now comes the practical question: where do you actually buy everything? After building dozens of headphones over the years, I've learned which suppliers are reliable, which ones to avoid, and how to find components that aren't always obvious. Let me save you the trial and error I went through.

The Complete Parts List

Here's everything you need for a complete headphone build:

Essential components: Drivers (2), ear cups/housings (2), ear pads (1 pair), headband, cable, connectors (if detachable), damping materials (foam, felt, cotton), and hardware (screws, nuts, washers, heat-set inserts).

Optional but recommended: Adjustment mechanism (slider or extending yokes), headband padding, cable management clips and strain relief, and protective grilles for open-backs.

Tools you'll need: Soldering iron, multimeter, precision screwdriver set, hot air gun or soldering iron tip for heat-set inserts, wire strippers, helping hands or small vise, and access to a 3D printer if printing your own cups.

Drivers: Your Main Suppliers

Parts Express (partsexpress.com): The go-to source for North American DIY builders: reliable quality control, reasonable prices, fast shipping. One caution about the house brand: most of the Dayton Audio catalog is loudspeaker drivers, not headphone drivers. The one Dayton part that genuinely belongs in a headphone build is the CE38MB-32 (part # 285-131), a 38 mm, 32-ohm mini driver around $3.49 that Dayton actually markets as a headphone replacement driver, and that the DIY community has real build history with. Its known character: bass-heavy in a closed-back design, so plan your damping accordingly.

Madisound (madisound.com): Another excellent US supplier. More focused on high-end components, sometimes carries drivers you won't find elsewhere. Good customer service and technical knowledge.

AliExpress: As we discussed in Driver Selection Guide (Chapter 7), proceed with caution. Use only for experimental builds with verified driver models.

Ear Pads

Ear pad sourcing depends on your design. If you're using an existing headphone chassis or headband, look for replacement pads compatible with that model. If you're designing from scratch, you have more options.

Amazon: Large selection of generic replacement pads. Search for "replacement ear pads" with your desired diameter (usually 90–105mm inner diameter for over-ear). Quality varies. Read reviews carefully.

Geekria (geekria.com): Specializes in replacement audio accessories. Better quality than random Amazon generics, reasonable prices, good selection of materials.

Brainwavz (brainwavzaudio.com): Well-regarded in the DIY community for quality pads. Earpads are their specialty. Higher price but consistent quality.

What to look for in pads: Inner diameter (must match your design), material (velour for breathability, leather/pleather for isolation), depth/thickness (affects driver-to-ear distance and sound), and attachment method (many DIY designs use a simple friction ring).

Headband Components

Headband components are where DIY gets interesting. For your first build, you have a few options:

Salvage from broken headphones: The headband and adjustment mechanism are often the hardest parts to design from scratch. A broken headphone with good mechanical components (even if the drivers are blown) is a valuable resource. Thrift stores and eBay are excellent sources.

3D print your own: Once you have CAD skills (covered in 3D Design for Headphones (Chapter 14)), you can print adjustment sliders, cups, and other structural components. Use PETG for structural parts: PLA is too brittle for parts that see repeated stress. If you're deciding *which* printer to buy, or whether you need one at all, Choosing a 3D Printer for Headphones breaks down what actually matters for these small parts: the machine recommendations stay current there rather than here.

Headband kits: A few suppliers offer headband components. Search "headphone headband parts" on eBay and Amazon. Quality varies dramatically.

Cables and Connectors

Covered in detail in Cables, Connectors & Hardware (Chapter 12), but briefly:

Wire: Mogami, Canare, or Belden from professional audio suppliers like B&H Photo, Sweetwater, or Redco Audio. For budget builds, 24 AWG stranded copper from Parts Express works fine.

Connectors: Neutrik for quality (3.5mm TRS jacks, mini-XLR). Amazon for budget 3.5mm jacks: buy name brands like Amphenol, not the cheapest option.

Damping Materials

Covered thoroughly in Damping Materials (Chapter 11), but the quick sourcing guide:

Acoustic foam: Parts Express carries Acoustipack and similar materials. Amazon has many options: search for "acoustic foam packing material" rather than "soundproofing foam" (different products).

Felt: Craft stores (Michaels, Hobby Lobby, Jo-Ann Fabrics). Adhesive-backed felt is convenient. Regular felt works fine too.

Cotton: Craft stores. Polyester fiberfill from a fabric store also works well for rear-chamber fill.

Hardware and Fasteners

McMaster-Carr (mcmaster.com): The absolute best source for hardware. They have every fastener, grommet, spring, and specialty part you could want, with detailed specifications. Fast shipping. Slightly higher prices but worth it for specific parts you can't find elsewhere.

Amazon: Good for assortment kits of common sizes (M3 screws in various lengths, heat-set inserts). Often cheaper than buying individually for common sizes.

Home Depot / Lowe's: Limited selection but immediately available. Good for standard screws when you need them today.

Heat-set inserts specifically: Search "M3 heat set inserts" on Amazon. A kit of 100 mixed sizes runs $10–15 and will last you many builds. These are essential for creating durable threads in 3D printed parts.

Measurement Equipment

Not strictly a component, but you'll want it for tuning. The full breakdown, what to buy at each budget tier, and what each rig can and can't tell you, lives in Budget Measurement Setup (Chapter 19). The gear recommendations stay current there rather than here.

One sourcing note worth making in this chapter: the usual entry-level calibrated USB mic, the MiniDSP UMIK-1, is frequently supply-constrained. Don't plan around a single source: any calibrated measurement microphone with a downloadable calibration file does the same job in REW, so keep an alternative in mind for when the UMIK-1 is out of stock.

You don't need measurement equipment for your first build: listening is sufficient. But once you're serious about tuning, measurement tools become invaluable.

Budget Planning

For a realistic first build budget:

- Drivers: $15–30
- Ear pads: $15–25
- Cable and connectors: $10–20
- Damping materials: $10–15
- Hardware (screws, inserts): $10–20
- 3D printing filament: $5–10 (if printing your own parts)

- **Total: $65–120**

Budget separately for measurement equipment if you want it: Budget Measurement Setup `(Chapter 19)` breaks down what each tier costs.

What's Next

The acoustic chamber is the most critical physical design element, and it's the next thing to get right. In Acoustic Chambers and Enclosures `(Chapter 9)`, we'll cover how chamber design affects sound, what materials work best, and how to design enclosures that actually sound good.

CHAPTER 9

Acoustic Chambers and Enclosures

Learn how acoustic chamber design affects headphone sound: volume, shape, materials, and how to design enclosures that actually work.

8 min read   makerphones.com/learn/acoustic-chambers-and-enclosures

You can have the best drivers in the world, but put them in the wrong enclosure and they'll sound terrible. The acoustic chamber, that space between the driver and your ear, and the volume behind the driver, is just as important as the driver itself. After designing dozens of headphone enclosures over the years, I've learned what works, what doesn't, and why. Let me share what I've figured out so you don't have to learn it all the hard way.

Why the Chamber Matters

As we discussed in How Headphones Create Sound (Chapter 1), you're building a complete acoustic system, not a driver housing. The chamber affects:

- **Bass extension and level:** Chamber volume is directly related to how deep your bass goes and how loud it is
- **Resonances:** Every enclosed volume has resonant frequencies, peaks and dips that color the sound
- **Driver behavior:** The back pressure in a closed chamber affects the driver's compliance and resonant frequency
- **Damping requirements:** Chamber shape and volume determine where and how much damping you need

Front Chamber vs. Rear Chamber

When you install a driver in a cup, it creates two separate acoustic spaces:

The front chamber is the space between the driver face and your ear, essentially the volume created by the ear cup and ear pad when they're on your head. The sound you hear is created here. Its volume (determined largely by the ear pad depth) affects the relative bass level and the sound's "openness."

The rear chamber is the enclosed space behind the driver (for closed-back designs) or the open space connected to the outside (for open-back). In closed-back designs, this volume is critical for bass tuning.

Open-back designs have a "rear chamber" that's connected to outside air through vents or openings. This eliminates rear-chamber resonances but also removes the bass reinforcement a sealed chamber provides.

Volume and Bass Response

For closed-back headphones, rear chamber volume directly affects bass extension. Here's the relationship:

Larger volume = lower bass resonance = better bass extension. But there are practical limits: larger cups are heavier and less comfortable.

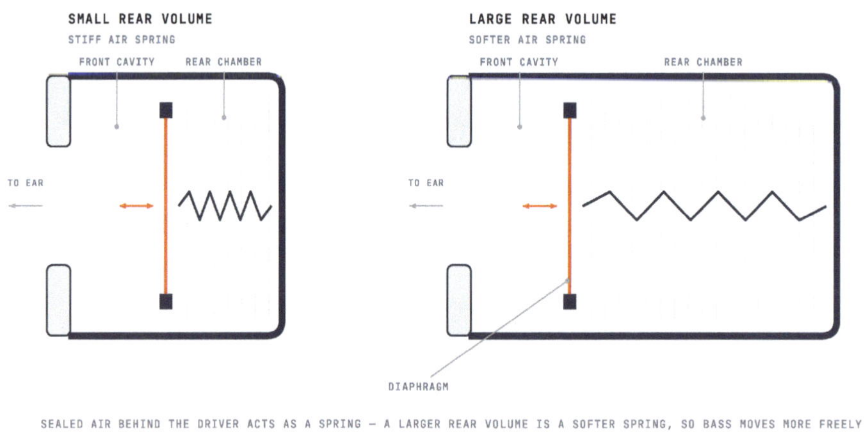

Fig. 9.1 — The rear-chamber air spring: smaller volume, stiffer spring.

The sweet spot for most closed-back over-ear headphone rear chambers is 80–150cc (cubic centimeters). Commercial headphones cluster here because it provides decent bass extension while keeping the cups manageable.

For open-back designs, this relationship is less direct because the rear isn't sealed. Open-back cups typically have 50–100cc of enclosed volume, but the open sections reduce the effective acoustic volume significantly.

Front Chamber Volume and Ear Pad Depth

The front chamber, the space between driver and ear, is primarily determined by your ear pad depth and inner diameter. A deeper pad puts more distance between driver and ear.

Push the driver farther from your ear and a few things happen together:

- Reduces high-frequency energy (highs fall off with distance)
- Creates a slightly larger front chamber volume (more bass, more spacious sound)
- Can improve soundstage (more "room" for the sound)

Pull it closer and you get the reverse:

- More high-frequency energy
- More intimate/direct sound
- Can sound bright if the driver has a lot of treble to begin with

Standard over-ear pads are typically 18–25mm deep. This range works well for most drivers. Going much shallower can make treble harsh; going much deeper can make bass boomy or sound lose focus.

Chamber Shape: Why Asymmetry Helps

A perfectly cylindrical chamber with parallel walls creates standing waves at specific frequencies. These show up as peaks in your frequency response and usually sound unpleasant, often described as a "boxiness" or "cupped hands" character.

Commercial headphone manufacturers use several strategies to minimize standing wave problems:

Angled walls: If the walls aren't parallel, standing waves can't develop as easily. A slight angled taper (even 5–10°) helps.

Irregular shapes: Oval or teardrop cup shapes have no two parallel surfaces, reducing standing wave problems.

Damping materials: Strategic placement of acoustic foam breaks up standing waves. This is why damping `(Chapter 11)` is such an important tool.

For DIY builders: don't design a perfect cylinder if you can avoid it. A slight oval or tapered shape costs you nothing in design complexity and helps acoustically.

Materials and Their Acoustic Properties

The material your cup is made from affects the sound. Hard, rigid materials reflect sound internally; softer or damped materials absorb it. Here's what to know about common DIY materials:

PLA (3D printing): Rigid, resonant. Good for structural parts but the material itself can ring. Adding internal damping materials compensates. Brittle. Consider PETG for structural pieces.

PETG (3D printing): Slightly more flexible than PLA, less resonant. Better choice for structural components like headbands. Similar acoustic properties to PLA for cups.

Wood: Naturally damped compared to plastic. Wood cups have a reputation for "warmth" that may partly be acoustic (internal damping) and partly the resonant character of wood itself. Harder to work with but worth exploring later.

Aluminum/metal: Very rigid, can be very resonant if not treated. Metal cups require careful damping treatment. More challenging for DIY fabrication.

Porting and Venting

Small ports or vents in the cup can dramatically change the acoustic character. This is how many commercial headphones achieve their specific bass tuning.

Rear chamber ports (closed-back): Adding a small port to the rear chamber converts it from sealed to vented. This lowers the bass resonance frequency and can extend bass, but also reduces bass level at the resonant frequency. Getting the port size right is tricky: small changes have big effects.

Front baffle vents: Some headphones have small vents in the front baffle (the plate the driver mounts to). These can reduce the "sealed" pressure against the ear and affect midrange response.

For your first build, avoid porting. It adds complexity without clear benefit when you're still learning. Master sealed design first, then experiment with ports later.

Practical Design Guidelines

Here's what I recommend for a first closed-back build:

- **Rear chamber volume:** 100–120cc (roughly a 7cm x 4cm x 4cm space as a reference)
- **Shape:** Slightly oval, avoid parallel walls where possible
- **Wall thickness:** Minimum 3mm for 3D printed parts (4–5mm better)
- **Material:** PETG for durability (PLA for prototyping)
- **No ports:** Keep it sealed for your first build
- **Damping plan:** Leave room for foam placement in design, don't fill the chamber with structure

For a first open-back build:

- **Open area:** 30–50% of rear surface area should be open (grille or vents)
- **Enclosed volume behind driver:** 50–80cc
- **Minimal damping needed in rear**: you won't have the sealed-chamber resonance problems

What's Next

The chamber ends where it meets your head, at the interface that does as much for the sound as anything inside the cup: ear pads. In Ear Pads and Comfort `(Chapter 10)`, we'll cover how pad choice affects both sound and comfort, and how to select the right pads for your build.

CHAPTER 10

Ear Pads and Comfort

Learn how ear pads affect both sound quality and comfort: material selection, thickness, attachment methods, and how to choose the right pads for your build.

8 min read   makerphones.com/learn/ear-pads-and-comfort

Ear pads are one of those components that seems minor until you realize they affect both how your headphones sound and whether you can actually wear them for more than 20 minutes. I've seen beautifully designed headphones fail because the pads were wrong. Let me explain what pads actually do, what the options are, and how to choose.

What Ear Pads Actually Do

Most people think of ear pads as purely a comfort feature. They're actually doing several acoustic jobs simultaneously:

Creating the front chamber: As we covered in Acoustic Chambers and Enclosures (Chapter 9), the space between your ear and the driver is the front acoustic chamber. The pad depth determines this distance, which affects high-frequency response and soundstage character.

Providing the seal: A good pad seal is essential for proper bass response. A leaky seal, one that lets air escape, reduces bass significantly. The pad material and shape determines how well it seals against different head shapes.

Absorbing sound: Pad material absorbs certain frequencies, affecting the final sound. Velour pads absorb more high frequencies than smooth leather/pleather pads, which is why the same headphone can sound different with different pad materials.

Affecting pressure distribution: How the pad contacts your head and ear distributes the clamping force. Good pad design spreads the force comfortably;

bad pad design creates pressure points.

Pad Materials

Velour / Fabric

Velour pads feel soft and comfortable, breathe well, and are generally the most comfortable for long listening sessions. The porous material absorbs some high-frequency energy, which often makes headphones sound slightly warmer or smoother with velour vs. leather.

Good for: Long listening sessions, warm climates, listeners who prefer smooth/relaxed sound signature

Watch out for: Worse isolation than leather, can wear out faster, absorb moisture and odors over time

Leather and Pleather (Protein Leather, Synthetic Leather)

Leather and pleather pads reflect more high-frequency energy than velour, which can make headphones sound brighter and more detailed. They seal better due to the smooth, non-porous surface. Real leather breathes slightly; pleather does not.

Good for: Better isolation, brighter sound character, easier to clean

Watch out for: Hot and sweaty in warm conditions, can cause discomfort during long sessions from heat buildup, pleather peels eventually

Memory Foam vs. Standard Foam

The foam inside the pad also matters. Memory foam conforms to your head shape, spreading pressure more evenly. This is often significantly more comfortable for longer sessions. Standard foam is fine but may create pressure points.

Memory foam pads cost more but are worth it for headphones you'll wear for hours at a time.

Hybrid Pads

Some pads use leather on the outer edge (for isolation) and velour or perforated leather on the inner surface (facing your ear, for breathability). These aim to get the sealing benefits of leather with some of the comfort benefits of velour. They can work well, though they're more expensive.

Pad Geometry: What the Numbers Mean

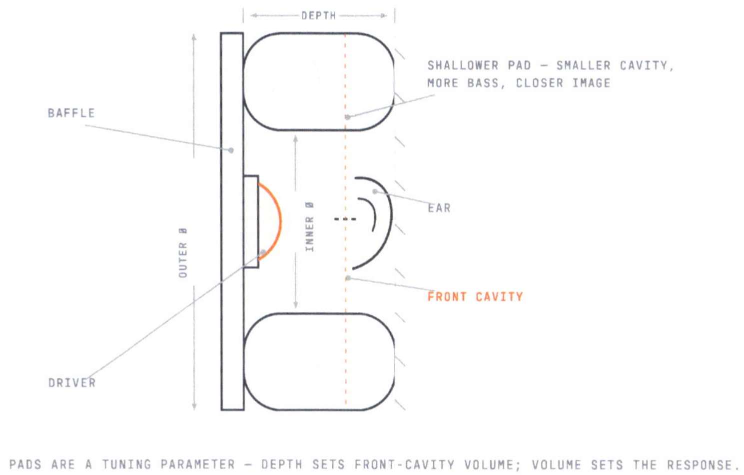

Fig. 10.1 — An ear pad in cross-section: inner diameter, depth, and outer diameter, and how pad depth sets the front-cavity volume between driver and ear.

Inner Diameter

The inner diameter is the opening size: the hole your ear goes into (or presses against, for on-ear designs). For over-ear headphones, this needs to be large enough to fit your ear comfortably. Most adults need at least 60mm inner diameter; 65–75mm is the comfortable range.

For your design, match the inner diameter to your ear cup opening. Standard over-ear headphone inner diameters are 90–105mm (the outside of the pad), with inner openings of 60–75mm.

Depth / Thickness

Pad depth is the distance from the flat surface that mounts to the cup to the flat surface that contacts your head. Typical over-ear pads are 18–30mm deep.

As we covered in the acoustic chambers chapter, depth affects driver-to-ear distance:

- Shallower pads (18–20mm): Driver closer to ear, brighter sound, more detail, less soundstage depth
- Medium pads (22–25mm): Good balance for most applications
- Deep pads (28–30mm): Driver farther from ear, often more spacious/airy sound, slightly reduced treble

Outer Diameter

The outer diameter determines the overall size of the pad. This needs to match your cup design: too small and the pad doesn't cover the cup edge properly; too large and it looks wrong and may not seal correctly.

Attachment Methods

Pads attach to cups in several ways. This matters for both your cup design and pad replacement later.

Friction ring / tension ring: A ring inside the pad that stretches over a lip on the cup. Very common on commercial headphones. Easy to swap pads, no tools needed. Requires designing a lip into your cup.

Adhesive: Pads with peel-and-stick adhesive on the back. Simple to attach, harder to replace. Good for prototyping or if you want to lock in a specific pad choice.

Screws / bolt-through: Some designs screw the pad frame directly to the cup. Very secure, allows easy removal with tools.

Magnetic attachment: Used in some premium commercial headphones. Complex to design for DIY.

For DIY builds: design a tension ring lip into your cup. It's the most user-friendly system and lets you experiment with different pads during tuning.

How Pad Changes Affect Sound

This is something I want you to internalize: changing pads can dramatically change how your headphones sound. This is both a feature (pad swapping is a powerful tuning tool) and a source of confusion for new builders.

When you change pad material from velour to leather, you'll typically hear:

- More treble presence (leather reflects high frequencies back toward the driver)
- Slightly different soundstage character

- Better bass (if seal improves) or similar bass (if seal was already good)

When you change pad depth (deeper pads):

- More spacious/airy sound
- Slightly reduced treble energy
- Sometimes more bass (larger front chamber)

This is why the T50RP modding community (see Design Methodology (Chapter 13)) has spent years trying different pad combinations. It's a significant and accessible tuning tool.

Comfort Engineering

Sound quality doesn't matter if you can't wear the headphones for long. Here's what makes pads comfortable for extended use:

Memory foam vs. standard foam: Memory foam wins for comfort, especially for sessions over 1 hour. If budget allows, use memory foam.

Breathability: If you run warm or live in a warm climate, velour or perforated leather significantly improves wearability. Sealed leather pads get hot fast.

Clamping force interaction: The pad and clamping force work together. Softer pads feel more comfortable even at higher clamping forces. This gives you some design flexibility: slightly higher clamp (better seal) can work if the pad is very soft.

Weight consideration: Pads can be surprisingly heavy. Check the weight before committing to them as part of an already-heavy design.

Recommended Pads for First Builds

Rather than chasing exotic options, here are reliable choices that work well:

Budget starter: Generic replacement pads from Amazon, search "90mm headphone pads velour" or "replacement ear cushions 90mm." $10–20. Hit or miss quality but fine for prototyping.

Reliable mid-range: Brainwavz HM5 pads or similar. Well-regarded in the DIY community. ~$25–35. Available in velour, leather, and hybrid versions. Standard 90mm size fits most designs.

Premium option: Dekoni Audio pads. Made for specific commercial headphones but also available in generic sizes. $40–70. Excellent quality and comfort if you want the best.

What's Next

With pads understood, the next component category is one of your most powerful acoustic tuning tools: damping materials. In Damping Materials (Chapter 11), we'll cover what materials work, where to place them, and how to use damping to achieve the sound you want.

CHAPTER 11

Damping Materials

Learn which damping materials work best for headphones, where to place them, and how much to use, the key to smooth, natural sound.

7 min read   makerphones.com/learn/damping-materials

Damping is one of those topics where a little knowledge makes a huge difference. I've seen beautifully designed headphones sound terrible because the builder didn't understand damping, and I've seen mediocre designs salvaged by smart damping choices. Here's what works, what doesn't, and how to use these materials strategically.

Why Damping Matters

When sound bounces around inside a headphone cup, it doesn't just disappear: it creates reflections that color the frequency response. Without any damping, you'd hear the direct sound from the driver plus all those reflections, which creates peaks and resonances that make things sound unnatural.

Damping materials absorb some of this reflected energy. Different materials absorb different frequency ranges with different efficiency. Your job as a builder is to figure out which frequencies need controlling and where to put the damping to address them.

The Main Damping Materials

Open-Cell Acoustic Foam

This is your primary workhorse. Open-cell foam lets air pass through it, which creates friction that absorbs sound energy, particularly in the high-frequency range.

The key spec is PPI (pores per inch). Higher PPI = smaller pores = more absorption of higher frequencies.

- **10 PPI foam:** Mostly bass/low-mid absorption, minimal high-frequency effect
- **20–30 PPI foam:** Good mid-range absorption, the most versatile choice
- **45–60 PPI foam:** Heavy high-frequency absorption, can sound "dead" if overdone

For most applications, 20–30 PPI foam is your starting point. Acoustipack from Parts Express is a well-regarded option. Generic acoustic foam from Amazon works too: look for open-cell foam, not egg crate style (which is mostly for diffusion, not absorption).

Felt

Felt is different from acoustic foam: it's denser and absorbs mid-to-high frequencies quite aggressively. A small piece of felt can make a big difference, which means it's also easy to overuse.

Felt is particularly useful for controlling specific resonance peaks. Place it over the driver face (lightly) to tame harsh treble, or use it in specific problem areas of the chamber. Start with thin felt (1–2mm) and work up.

Craft store felt works fine. Get both regular and adhesive-backed varieties: the adhesive version makes positioning easier during experimentation.

Cotton / Fiberfill

Cotton batting or polyester fiberfill (the stuff inside stuffed animals) is commonly used to fill rear chambers. It has a useful acoustic property: it effectively makes the air in the chamber "behave like" a larger volume, which extends bass response.

This is one of those things that sounds like audiophile nonsense until you measure it. A chamber filled with fiberfill genuinely measures differently than an empty chamber of the same size: the bass extends lower and the response is often smoother.

Use loosely. Don't pack it tight. Think of it as lightly filling the space.

Viscous Materials (Dynamat, Blu-Tack)

Dynamat and similar automotive vibration-damping materials can be applied to the inside walls of cups. They add mass and damping to the cup structure itself, reducing resonance in the walls.

This matters more with thin, lightweight cup materials. If your cup walls are vibrating (you can sometimes feel this), Dynamat helps. For typical 3D printed cups with 3–5mm walls, this is less necessary.

Blu-Tack (the stuff used to hang posters) is a cheap alternative for quick experiments. Easy to apply and remove, reasonably effective.

Where to Place Damping

Location matters as much as material. Here's a general placement guide:

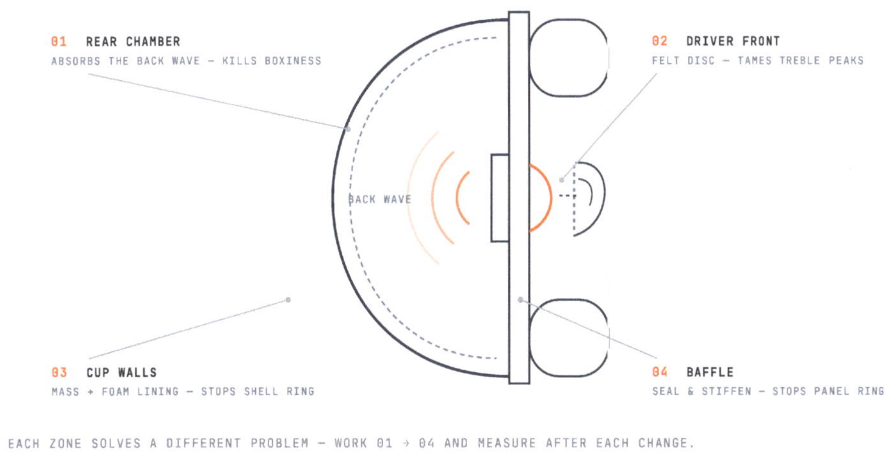

Fig. 11.1 – The four damping placement zones, rear chamber, driver front, cup walls, and baffle, each solving a different problem.

Rear Chamber (Behind the Driver)

This is the most impactful location for closed-back headphones. Adding foam here primarily affects bass and lower midrange, controls rear-chamber resonances, and generally makes the sound less "hollow" or "boxy."

Approach: Line the walls of the rear chamber with a thin layer of foam (don't cover every surface, start with one wall). Add fiberfill loosely if you want to extend bass. Test after each addition.

Driver Front (Between Driver and Ear)

This is the most sensitive location. Materials here directly affect what you hear, particularly in the midrange and treble. A tiny piece of thin foam or felt over the driver can dramatically change the sound.

Approach: Only use very thin, light material here. Perforated material (foam with holes, thin fabric) is gentler. Start with nothing and add only if you have specific treble problems. This is a last resort, not a starting point.

Cup Walls

Lining the inner walls of the cup controls wall reflections and reduces cup resonances. Medium-weight felt or foam works well here.

Approach: If the cup has a "boxy" character, line the walls. Leave the front and rear open initially. Test with thin material first.

Baffle (The Plate the Driver Mounts To)

The baffle can have its own resonances. A thin ring of foam or felt around the driver, between the driver and baffle, can reduce baffle reflections.

The Tuning Process

Here's the process I use when tuning with damping:

1. **Start bare:** Do your first listening test with no damping. Note the problems: harsh treble? Boomy bass? Hollow midrange?
2. **Address the biggest problem first.** One change at a time.
3. **Use measurement if available.** Frequency response graphs show you exactly where the problems are. See Budget Measurement Setup `(Chapter 19)`.
4. **Start with less than you think you need.** Overdamping is a real problem: headphones can sound dead and lifeless if you absorb too much energy.
5. **Listen critically after each change.** Put them on, listen to familiar music for 2–3 minutes. What changed? Better? Worse? Different?
6. **Trust your ears.** If it sounds good, it IS good, regardless of measurements.

Common Mistakes

> ⚠ CAUTION
>
> **Mistake 1: Filling the entire chamber with foam.** Problem: Kills dynamics, sounds dead. Solution: Place foam strategically instead of filling the whole chamber.
>
> **Mistake 2: Using only cotton fill.** Problem: Unpredictable, can shift, inconsistent absorption. Solution: Use foam/felt for main damping, cotton for fine-tuning only.
>
> **Mistake 3: Damping front of driver first.** Problem: Very sensitive, easy to overdo. Solution: Exhaust all other options before damping driver front.
>
> **Mistake 4: Not testing between changes.** Problem: Add 5 things at once, can't tell what helped. Solution: One change at a time, test each.
>
> **Mistake 5: Assuming more is better.** Problem: Overdamped headphones sound lifeless. Solution: Start minimal, add only what's needed.

What's Next

Next up is the final component category: cables, connectors, and the small hardware that holds everything together. In Cables, Connectors & Hardware (Chapter 12), we'll cover wiring, connection options, and all the small parts that hold your headphones together.

CHAPTER 12

Cables, Connectors, and Hardware

Learn which cables, connectors, and hardware actually matter for DIY headphones, and which expensive options you can skip.

10 min read   makerphones.com/learn/cables-connectors-hardware

You can have the perfect drivers, the ideal acoustic chamber, and beautifully tuned damping - but if a cable breaks or a connector fails, your headphones are useless. I learned this the hard way at HeadRoom when we'd see beautifully sounding headphones come back for repair because of a $0.50 connector that failed. This chapter covers which cables, connectors, and hardware actually matter.

What Actually Matters in Cables (And What Doesn't)

Let's start with the controversial bit: **for headphone cables carrying analog audio at normal listening levels, the wire itself makes virtually no audible difference beyond basic electrical properties.**

I've done blind listening tests with $5 cables and $500 cables at HeadRoom. With properly matched levels, nobody could hear a difference. Not me, not our trained listeners, not visiting audiophiles. The expensive cables felt nicer, looked better, and sometimes lasted longer - but they didn't sound different.

What DOES matter: proper gauge (thickness) for the signal and impedance; adequate shielding to reject noise and interference; mechanical durability: flex life, strain relief, jacket material; connector quality: secure connection, low resistance contacts.

What usually DOESN'T matter audibly: exotic conductor materials (oxygen-free copper vs. regular copper), cable geometry beyond basic twisted-pair design, "burn-in" claims, directional arrows on cables, or cryogenic treatment.

Spend your money on better drivers (Chapter 7) or ear pads (Chapter 10) instead of exotic cables.

Wire Gauge Selection

Wire gauge matters for two reasons: electrical resistance and mechanical durability.

Understanding AWG (American Wire Gauge)

Wire thickness is measured in AWG. **Smaller numbers = thicker wire.** Yes, it's backwards.

- **24 AWG:** Standard for most headphone cables, good for 3-10 feet
- **26 AWG:** Thinner, more flexible, fine for short runs (under 5 feet)
- **28 AWG:** Very thin, flexible, use only for very short runs or low-power applications
- **22 AWG:** Thicker, stiffer, only needed for very long runs or special cases

For DIY headphones: 24 AWG is your sweet spot. Flexible enough to work with, thick enough to be durable, low enough resistance for any headphone impedance.

Resistance Matters (A Little)

Cable resistance creates a voltage divider with your headphone impedance (Chapter 3). For typical headphones, cable resistance under 1Ω is fine for low-impedance headphones, under 2Ω for medium impedance. 24 AWG copper wire has about 0.025Ω per foot. Even a 10-foot cable is negligible compared to headphone impedance.

Shielding: When You Need It

Shielding blocks electromagnetic interference (EMI) from getting into your audio signal.

You need shielding if: your cable runs near power cables or transformers; you're in an electrically noisy environment; your cable is longer than 6 feet; or you're building portable headphones near phone/computer RF.

You probably don't need shielding if: the cable is short (under 4 feet), you're in a clean electrical environment for home use, or you have high-impedance headphones (less susceptible to noise).

Shielding types: Braided shield (woven copper braid, excellent coverage, flexible) is the best balance of flexibility and protection for DIY builds.

Connector Options: From Simple to Sophisticated

Connectors are where most DIY builds fail. A broken connector makes perfect headphones useless. Let's go through your options.

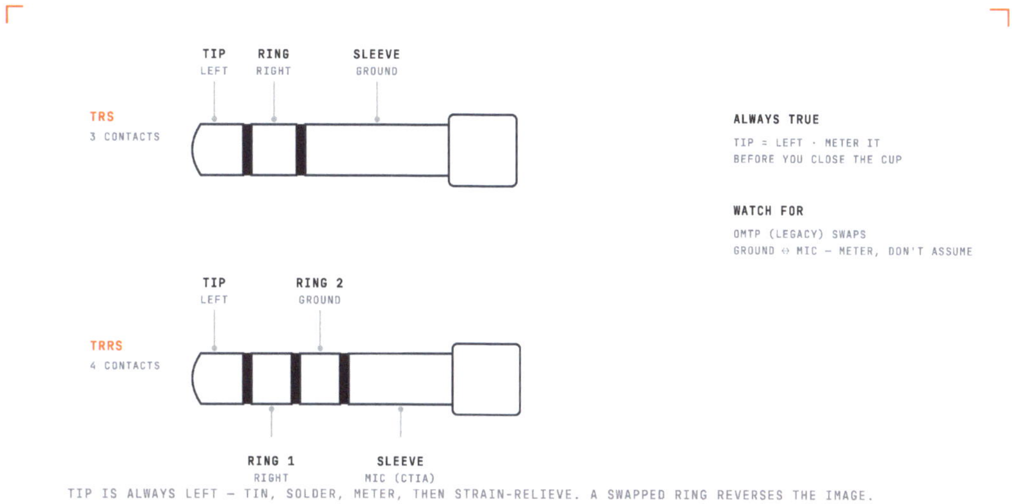

Fig. 12.1 — TRS and TRRS contact layout: tip is always left; the conductor map is the wiring step builders most often get wrong.

Option 1: Hardwired (No Connector)

Cable permanently attached. Pros: simplest to build, one less failure point, lowest cost, best electrical connection. Cons: can't swap cables, cable damage means repair or trash. Best for: simple builds, prototypes.

Option 2: 3.5mm TRS Jack (Standard Headphone Jack)

Pros: universal cables available everywhere, cheap replacement cables, reliable when done well. Cons: larger connector housing (design consideration), can be fragile if cheaply made. Technical note: TRS = Tip, Ring, Sleeve. Tip = Left, Ring = Right, Sleeve = Ground.

Option 3: 2.5mm TRS or TRRS🔗

Smaller connector housing, can do balanced with TRRS (4-pole). Fewer cable options and more fragile than 3.5mm. Best for smaller/lighter builds where size matters.

Option 4: Mini-XLR (Professional Detachable)🔗

Very secure locking connection, professional appearance, takes abuse without failing. More expensive ($5-15 per connector), requires specific housing design, less universal cable availability. Best for serious builds where durability matters most.

Option 5: MMCX (Micro Miniature Coaxial)🔗

Tiny footprint, rotates freely (good for cable management), many aftermarket cables available. Can wear out with repeated insertion. Best for IEMs and very small/light headphones.

My Recommendation for DIY Builds🔗

First build: Hardwired. Get it working, learn the process, don't add connector complexity.

Second build: 3.5mm TRS jack. Learn connector integration, enjoy cable swapping, use standard cables.

Advanced builds: Mini-XLR if you want professional quality and don't mind the cost.

Hardware: The Little Parts That Matter 🔗

Screws and Fasteners🔗

For 3D printed parts: Heat-set inserts (brass threaded inserts) are the best option: install with a soldering iron, create durable metal threads in plastic. Self-tapping screws work but wear out plastic threads over time.

Recommended sizes: M3 x 0.5mm pitch for most headphone assembly; M4 x 0.7mm pitch for larger/heavier components; M2 x 0.4mm for small detail work.

Length selection: Screw should engage at least 1.5x its diameter in thread. So an M3 screw needs 4.5mm of thread engagement minimum.

Headband Hardware🔗

Adjustment mechanisms: Ratchet/detent system clicks into position and stays put; friction fit is simple and reliable but can slip; spring-loaded pins are secure and adjustable but require more complex design.

Materials for headband structure: PETG (3D printed flexible plastic) is better than PLA. PLA fatigues and breaks. Steel spring bands are durable but can be uncomfortable. Aluminum is light and strong but expensive to fabricate.

Ear Cup Pivots and Hinges

Ear cups need to rotate to conform to your head shape. For DIY: simple screw pivot works fine if you use a wave washer for tension and a drop of threadlocker.

Cable Routing and Strain Relief

Cable failures almost always happen at the entry point to the ear cup. Strain relief is critical.

Good strain relief methods: Rubber grommet (simple, effective, available in many sizes); integrated designed relief (built into your ear cup, no extra parts); heat shrink tubing (adds flexibility transition, prevents sharp bends); commercial strain relief boots (professional look, reliable).

Practical Sourcing Guide

Wire and cable, Budget option: Parts Express (24 AWG speaker wire, $0.15-0.30/foot), Amazon ("headphone cable bulk wire"), Monoprice. **Better option:** Mogami 2893 (professional miniature quad cable, $2-3/foot), Canare L-4E6S (star quad, excellent noise rejection, $1-2/foot), Redco Audio.

Connectors, 3.5mm TRS jacks: Neutrik NMJ4HF (gold standard, $3-5 each), Amphenol KS3P (good quality, $1-2 each). **Mini-XLR:** Neutrik NC3MRX (3-pin right angle, $5-7 each). **MMCX:** Mouser/Digikey ($2-4 each, reliable).

Hardware, Heat-set inserts: McMaster-Carr (M3 brass inserts, $0.10-0.30 each) or Amazon kits ($10-15 for 100 pieces). **Screws:** McMaster-Carr for best selection; Amazon for assortment kits.

Assembly Considerations

Good soldering practices: Clean tip, proper temperature (650-700°F); tin both surfaces before joining; heat the joint, not the solder; use rosin-core solder (60/40 or 63/37); don't move joint while cooling.

Heat shrink tubing: Slide on BEFORE soldering (obvious but easy to forget); 2:1 shrink ratio minimum; heat evenly.

Testing before final assembly: Use multimeter on continuity mode: test each connection from connector to driver, test for shorts between channels. Then do a listening test with drivers in free air before final assembly.

Common Mistakes to Avoid

> ⚠ CAUTION
>
> **Mistake: Using wire too thin for the run length.** Problem: Audible high-frequency roll-off, weak bass. Solution: 24 AWG for runs up to 10 feet.
>
> **Mistake: No strain relief.** Problem: Cable breaks at cup entry within weeks. Solution: Always include proper strain relief, test flex cycles.
>
> **Mistake: Cheap connectors.** Problem: Intermittent connection, oxidation, failure. Solution: Buy quality connectors (Neutrik, Switchcraft). They last forever.
>
> **Mistake: Over-tightening screws in plastic.** Problem: Stripped threads, cracked plastic. Solution: Use heat-set inserts or stop when snug, not tight.
>
> **Mistake: No service loop.** Problem: Can't repair damaged connection without complete tear-down. Solution: Leave 2-3 inches extra cable inside cup.
>
> **Mistake: Reversing left/right channels.** Problem: Stereo image backwards (sounds weird). Solution: Mark your wires during assembly, test before closing up.
>
> **Mistake: Sharp edges on cable routing.** Problem: Cable insulation cut over time, short circuit. Solution: File/sand any edges cable touches, use grommets.

What's Next

With the components covered, it's time to put them together into an actual build. Start with Design Methodology `(Chapter 13)`, the complete workflow from concept to finished headphones, then continue through 3D design, chamber design, assembly, and damping strategy.

Or if you want to jump straight into building, check out Your First Build where we walk through a simple, achievable first project.

PART

3

Design & Build

CHAPTER 13

Design Methodology

The complete workflow from concept to finished headphones: how to design, prototype, test, and refine your DIY headphones systematically.

13 min read makerphones.com/learn/design-methodology

You now understand how headphones work, what components you need, and how they interact. But how do you actually go from "I want to build headphones" to holding a finished pair in your hands? That's what this chapter is about: the complete workflow from initial concept through testing and refinement. This is the design process I developed over 25+ years in professional audio, adapted for DIY builders.

The Five-Phase Design Process

Every successful headphone design follows the same basic path, whether you're building one pair or manufacturing thousands:

1. **Define Requirements**: What are you actually trying to build?
2. **Initial Design**: Create the basic structure and select components
3. **Prototype & Test**: Build it, measure it, listen to it
4. **Iterate & Refine**: Fix problems, improve performance
5. **Document & Finalize**: Lock down the design for repeatability

Most DIY builders skip straight to phase 2 or 3, which is why their first builds often fail or disappoint. Let's walk through each phase properly.

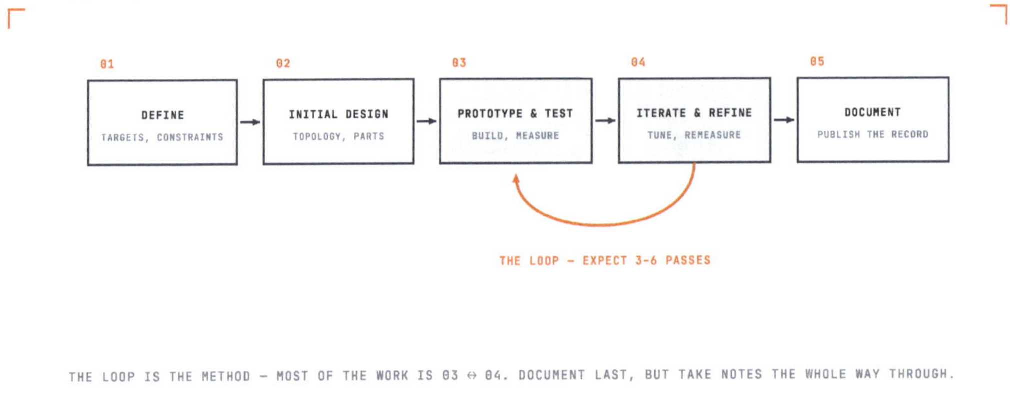

Fig. 13.1 – The five-phase design process. Most of the real work is the prototype-and-iterate loop in the middle.

Phase 1: Define Requirements

Before you design anything, answer these questions honestly. Write down your answers. They'll guide every decision.

What's Your Primary Use Case?

Critical listening / studio use: Needs flat frequency response (Chapter 2), low distortion, comfort for long sessions. Design direction: open-back (Chapter 4) for natural sound, larger chambers, quality drivers. Can sacrifice: portability, isolation.

Portable / commuting: Needs isolation, durability, comfortable fit, efficient drivers. Design direction: closed-back, lightweight, detachable cable. Can sacrifice: ultimate sound quality, large soundstage.

Gaming / multimedia: Needs comfort for hours, good imaging, possibly mic integration. Design direction: open or closed depending on environment, focus on fit.

Learning / experimentation: Needs simple assembly, low cost, instructive build process. Design direction: basic design, standard components, easy to modify.

What Are Your Constraints?

Budget: Under \$50 means salvaged or budget drivers, simple printed design, hardwired cable. \$50–150 gets good drivers, quality pads, basic detachable ectors. $150–300 supports premium drivers, multiple iterations, professional connectors.

Tools & equipment: Do you have a 3D printer or access to one? Can you solder reliably? Do you have measurement equipment (not required but helpful)? What's your CAD skill level?

Time: Quick weekend build means use existing designs, minimal customization. Month-long project supports original design, basic testing, some iteration. 3+ months allows full custom design, extensive testing, multiple prototypes.

What's Your Target Performance?

Be realistic. Your first DIY headphones won't beat $500 commercial models, but they can absolutely compete with $100–200 headphones if designed well.

Define your sound signature goals (neutral/reference using Harman curve, bass-emphasized, bright/detailed, or warm/smooth) and comfort requirements (maximum wearing time, weight limit, clamp force preference).

Document Your Requirements

Create a simple spec sheet. Here's an example:

Project: Open-Back Studio Headphones V1

- **Use case:** Home listening and mixing, 2–4 hour sessions
- **Sound target:** Neutral with slight bass lift (Harman-inspired)
- **Budget:** $150 total
- **Timeline:** 6 weeks from start to finished
- **Weight target:** Under 350g
- **Special requirements:** Must be 3D printable, detachable cable

Phase 2: Initial Design

Now you translate requirements into actual design decisions.

Component Selection

Drivers first: This is your most important decision. Reference Driver Selection Guide `(Chapter 7)` for detailed guidance. For a neutral open-back at $150 budget: 50mm dynamic driver, 32Ω impedance. Good candidates: Dayton Audio drivers. Budget allocation: $30–50 for driver pair.

Ear pads: Choose based on comfort needs and acoustic goals. See Ear Pads and Comfort `(Chapter 10)`. For a neutral open-back: velour pads for comfort, medium thickness (20–25mm). Budget: $15–25.

Cable & connectors: Reference Cables, Connectors, and Hardware `(Chapter 12)`. For this example: 3.5mm TRS jacks, quality cable. Budget: $15–20.

Damping materials: See Damping Materials `(Chapter 11)`. Start with basics: acoustic foam, felt, cotton batting for tuning. Budget: $10–15.

Hardware: Heat-set inserts, screws, headband components. Budget: $10–20.

Total component budget: $80–130, leaving $20–70 for iteration and mistakes.

Acoustic Chamber Design

Covered in depth in Acoustic Chamber Design `(Chapter 15)`, but here's the overview:

For open-back: Chamber volume 60–100cc; slightly oval shape to match ear; driver mounting 18–22mm from ear; rear opening 30–50% of rear surface area.

For closed-back: Chamber volume 80–120cc; non-parallel walls to reduce standing waves; plan damping placement locations; small controlled vents if needed for tuning.

Mechanical Design

Headband system: Adjustment range to accommodate head circumference 50–62cm; PETG for flexibility and durability (not PLA, it breaks); removable padding for cleaning/replacement.

Ear cup mounting: Pivot points allowing ±15° rotation minimum; plan for 1000+ adjustment cycles.

Assembly strategy: Screws recommended for first build; plan access for driver installation and wiring; ensure you can get back in to fix things.

Create CAD Models

See 3D Design for Headphones `(Chapter 14)` for detailed CAD guidance. Start simple: model the driver (import actual dimensions), create basic ear cup shape around driver, add mounting features, design headband attachment, add screw holes and heat-set insert locations.

Design for printing: Minimize overhangs (keep under 45° where possible); plan print orientation; wall thickness minimum 2–3mm for strength.

Phase 3: Prototype & Test

Now you actually print the thing. Your first prototype will have problems — mine always do — and that's the point: you want to find them fast.

Print Order Strategy

Don't print everything at once. You'll waste material when you need to make changes.

Print 1: Ear cup only (one side). Test driver fit, verify pad mounting, check connector clearance, test basic assembly. If this works, print the second cup. If not, redesign before printing more.

Print 2: Headband system. Test adjustment mechanism, verify range of motion, check strength.

Print 3: Full assembly after ear cups and headband validated.

Assembly Testing

Before installing drivers: Does everything fit together? Are heat-set inserts aligned? Do cups pivot smoothly? Does headband adjust easily? Any interference or binding?

Driver installation: Follow Driver Mounting and Assembly (Chapter 16). Wire one side first, test, then wire second side. Test continuity with multimeter. Check for shorts between channels.

Initial listening (before damping): Do both channels work? Rough sound balance (obviously broken or reasonable)? Any buzzing, rattling, or mechanical issues? Comfort check on your head.

Measurement (If Available)

If you have measurement capability (see Budget Measurement Setup (Chapter 19)): take 5 measurements per side, remove and reposition each time; average the measurements; note any obvious peaks or problems; save this as your "stock" reference.

Comfort Testing

Critical comfort checks: Can you wear them for 30+ minutes without discomfort? Hot spots or pressure points? Clamp force acceptable? Weight distribution even? Do they stay in place when you move?

If comfort fails, you must fix it before continuing. Sound quality doesn't matter if they're unwearable.

Phase 4: Iterate & Refine

Iteration is what separates a good design from a great one. Plan for multiple rounds.

Damping Tuning

Follow the process in Tuning with Damping `(Chapter 22)`. Start with back wave damping: add small amount of foam behind driver, measure and listen, add more if needed (but less is often more). Then address specific problems: peak around 5–7kHz means try felt on driver front (very light); muddy bass means check seal, may need less damping; hollow midrange may mean too much damping.

Iteration process: Baseline measurement → make ONE change → measure again → listen critically → keep change if better, remove if worse → document everything → repeat.

Mechanical Refinements

Common issues and fixes: Cups rotate too freely or too stiff: adjust wave washer tension or add friction material. Headband adjustment slips: increase detent depth or add friction material. Cables pull on cups: add strain relief or cable routing clips. Parts crack or break: increase wall thickness, add ribs, switch to PETG if using PLA. Squeaks or creaks: add small amount of lubricant to pivot points.

Aesthetic Refinements

Only after everything works and sounds good: surface finish (sanding, vapor smoothing, painting), logo or branding, color coordination with pads/cable. Don't waste time making broken headphones look pretty.

When to Stop Iterating

You could tweak forever. Here's when you're done: sound quality meets your target curve within 3–5dB with no obvious problems; comfort allows wearing for target duration without issues; no parts breaking or wearing out in testing; you can build another pair that sounds the same.

You can chase another half-dB forever. I don't. Once it hits the target and nothing's rattling, I lock the files and build the second pair.

Phase 5: Document & Finalize

This step separates a one-off prototype from a repeatable design.

Design Documentation

CAD files: Final versions with clear naming (v1.0, v2.0, etc.); export STLs with good naming; save print settings used; document material (PETG, PLA, print temp, etc.).

Bill of Materials (BOM): Every part with quantity; supplier links; part numbers or specifications; cost per unit; alternative sources if available.

Assembly instructions: Step-by-step with photos; special tools needed; torque specs for screws; wire routing diagram; damping material placement.

Performance Documentation

Measurements: Final frequency response graphs, left/right channel matching, distortion measurements if available, comparison to target curve.

Specifications: Impedance, sensitivity, weight, dimensions, cable length.

Build Log

Document your journey: what worked and what didn't; design iterations and why; mistakes made and lessons learned; time invested per phase; total cost breakdown.

This is valuable for you on the next build and invaluable if sharing the design.

Common Design Methodology Mistakes

⚠ CAUTION

Mistake: Skipping the requirements phase. Problem: Build something that doesn't meet your actual needs. Solution: Spend 30 minutes defining requirements before designing anything.

Mistake: Changing multiple things at once. Problem: Can't tell what change caused what effect. Solution: One change per iteration, measure each time.

Mistake: Not documenting changes. Problem: Forget what you did, can't recreate successes. Solution: Keep a build log, take photos, save measurements.

Mistake: Printing full assembly before testing. Problem: Waste material when design needs changes. Solution: Print and test incrementally.

Mistake: Focusing on aesthetics before function. Problem: Pretty headphones that sound bad or break. Solution: Make it work, then make it pretty.

Mistake: No comfort testing. Problem: Headphones that sound great but hurt to wear. Solution: Wear them for extended periods during development.

Mistake: Treating first prototype as final design. Problem: Missing opportunities for improvement. Solution: Plan for 2–3 iterations minimum.

Design Methodology for Different Builder Levels

Beginner (First Build)

Use an existing design or very simple original design. Focus on assembly and basic tuning. Measure if possible, but listening is okay. One iteration of damping tuning. Document what you learned. Goal: complete a working pair, learn the process.

Intermediate (Second-Third Build)

Define clear requirements. Original design or significant modification of existing. Measurement-guided tuning. 2–3 iterations. Complete documentation. Goal: create original design that meets specific requirements.

Advanced (Multiple Successful Builds)

Detailed requirements with target measurements. Complex original designs. Extensive measurement and modeling. Multiple prototypes with different approaches. Professional-level documentation. Consider sharing designs with community. Goal: designs that compete with commercial products, contribute to community knowledge.

What's Next

Each of the next chapters takes one piece of this process apart: 3D Design for Headphones (Chapter 14) (CAD skills and printability strategies); Acoustic Chamber Design (Chapter 15) (detailed geometry and acoustics); Driver Mounting and Assembly (Chapter 16) (physical assembly techniques); and Damping Strategy and Application (Chapter 17) (practical damping implementation).

Or jump to Your First Build for a guided walkthrough of a simple but complete project.

CHAPTER 14

3D Design for Headphones

How to model headphone parts in CAD and prepare them to 3D-print: parametric design, modeling around your driver, tolerances, and printability.

9 min read makerphones.com/learn/3d-design-for-headphones

The parts that make a headphone *yours*, the cups, the baffle, the headband slider, are exactly the parts nobody sells. A CAD model and a 3D printer are how you make them. CAD is where the design process stops being a plan and turns into something you can hold, seat a driver into, and put on your head.

Why model it yourself

You can buy drivers, pads, cable, and hardware. You can't buy the one cup that fits *your* driver, seals against *your* pad, and holds the air volume *your* design needs. That part is yours to make.

The good news: 3D printing makes it cheap to be wrong. A failed print costs an hour and a few cents of filament, so there's no reason to agonize over getting the first one perfect. Print it, find out what's off, change a number, print again. That loop is most of what designing a cup actually is.

Pick a tool you'll actually finish a part in

CAD tools split into two camps. *Parametric* tools (FreeCAD, Onshape, Fusion 360, SolidWorks) build geometry from dimensions you can edit later. *Direct* or *mesh* tools (Blender, Tinkercad) let you push shapes around freely. For headphones, parametric wins, and it isn't close: you will change the same dimension twenty times, and you want that to be easy.

Several cost nothing to start: FreeCAD is free and open source; Onshape and Fusion 360 have free or hobbyist tiers (check their current terms, which change). If you've never touched CAD, Tinkercad is the gentlest on-ramp: learn the basics there, then move to something parametric when you outgrow it. The tool matters less than committing to one long enough to get fluent.

Build it from named dimensions

Here's the habit that makes everything else faster: define the dimensions you care about as named parameters first: driver diameter, cup depth, wall thickness, pad inner diameter, then build geometry that references those parameters instead of typing raw numbers into the model.

Why bother? Because your first cup *will* be wrong. The chamber will be too small, or the pad mount the wrong depth. With a parametric model, fixing it means changing one value and watching the whole part update. With a model you "drew" by eye, it means starting over. This is the single biggest reason hobby headphone CAD goes faster than people expect, and the reason it connects so cleanly to the iterate step in the design process (Chapter 13).

> **TIP**
>
> Set your driver diameter and pad inner diameter as named parameters at the very top of the model, before you draw a single line. Almost every fit you care about traces back to those two numbers, and you'll change them the day you swap parts.

Design around the driver first

The driver is your fixed point; everything else mounts to it, so model it first. And measure your actual driver with calipers: outer diameter, mounting depth, any lip or flange. Datasheet drawings and the part in your hand drift apart more often than you'd like, and a bore that's half a millimeter off is the difference between a driver that seats and one that rattles.

From there you're shaping three things: the baffle or bore the driver seats into, the small cavity in front of it, and the volume behind it. Those front and rear volumes do real work on the sound, the chapter on acoustic chambers and enclosures (Chapter 9) covers what they do and why. Here, the job is making them buildable.

Think in printable pieces

A headphone isn't one part; it's an assembly: cup, baffle, pad mount, yoke, headband slider. Trying to model and print it as a single object fights the printer and makes every revision expensive. Break it into pieces that each print clean in their best orientation, and join them with inserts or press-fits.

This also lets you iterate one piece at a time: if the chamber's wrong, you reprint the cup, not the whole headphone. And it lets you mix made and bought parts: print the cup and baffle, salvage a headband from a broken pair (as the sourcing guide (Chapter 8) suggests), and bridge the two with a printed bracket.

Design for the printer

A model that looks right on screen can be impossible to print or impossible to fit together. Build these in from the start:

Wall thickness. Too thin and a cup wall warps as it cools and rings when the driver plays. For a structural cup, 2–3 mm of PETG is a sensible starting point, rigid and acoustically dead without wasting filament. Tune from there.

Tolerance and fit. Printed parts aren't dimensionally perfect, and they vary printer to printer. Any two parts that mate need a planned gap, a few tenths of a millimeter for a friction fit, and the only way to know the right number for *your* printer is to test it.

Threads and inserts. Don't print fine threads; they strip. For any joint you'll assemble and disassemble, design a boss sized for a heat-set insert (the kind in the sourcing guide (Chapter 8)). Brass inserts in plastic outlast printed threads many times over.

Orientation and supports. Design so the surfaces that matter, the sealing face, the pad-mount ring, print cleanly, without supports scarring them. If a critical surface needs supports, reorient the part or split it into two pieces instead.

Cable routing. Model the cable entry and strain relief into the part now. A hole you forgot becomes a drill job later, and drilling a finished print is how you crack it.

Print a coupon before the whole cup

Before you commit to an eight-hour cup print, print a small test coupon that exercises only the fits you care about: the driver bore, the insert boss, the pad-ring clearance. Five minutes of coupon saves an evening of reprints, and it's the physical version of testing a small thing before betting the build on it.

Common Mistakes

> ⚠ CAUTION
>
> - **Modeling parts to exact nominal size.** A 40.0 mm bore for a 40.0 mm driver gives you a part that won't seat, or one that's loose. Add clearance, tuned to your printer with test fits.
> - **Walls too thin to be dead.** A thin cup rings and colors the sound. Give structural walls real thickness.
> - **Printing threads.** They strip under assembly. Use heat-set inserts for anything you'll take apart.
> - **Supports on a sealing surface.** Support scars wreck a pad or driver seal. Reorient or split the part so critical faces print clean.
> - **Forgetting the cable.** Model the entry and strain relief from the start, not after the part is printed.

What's Next

You've got parts that fit and print. The next question is what those internal volumes actually do to the sound: the cavity in front of the driver, the chamber behind it, and how you vent them. Acoustic chambers and enclosures `(Chapter 9)` covers the fundamentals; a dedicated walk through designing a chamber for your own build is coming as this section fills out.

CHAPTER 15

Acoustic Chamber Design

Designing the front and rear chambers of your headphone, how volume, sealing, and venting shape bass and balance, and how to make those calls in your own build.

10 min read   makerphones.com/learn/acoustic-chamber-design

The driver gets all the attention, but the two pockets of air around it, one in front of the diaphragm, one behind, shape as much of the final sound as the driver does. Acoustic chambers and enclosures (Chapter 9) covered what those chambers are and why they exist. This chapter is about making the actual design calls for your build: what to set, what each setting changes, and how to leave yourself room to fix it when the first version isn't right.

Get the chambers wrong and no amount of damping rescues the build. You'll be sanding the edges off a problem that lives in the geometry. Get them right and most of the work's done before you tune anything.

The two chambers, and why the driver cares

Every over-ear headphone has a front chamber, the space between the diaphragm and your eardrum, set mostly by cup depth and pad thickness, and a rear chamber, the volume sealed or vented behind the driver. The diaphragm has to move air on both sides at once, so it's always working against both volumes simultaneously. Anything you do to either one changes how freely it moves, and therefore how the build sounds.

The most useful mental model is a spring. The air trapped behind the diaphragm pushes back when the diaphragm tries to move into it, exactly like a spring resisting compression. How stiff that spring is depends on how much air there is, and the stiffness of that spring is one of the biggest levers you have over the bass.

Rear volume sets your bass

A larger sealed rear chamber is a softer spring. It lets the diaphragm move more freely at low frequencies, which gives you deeper, more extended bass. A small sealed rear chamber is a stiff spring that fights the diaphragm on every excursion, producing bass that's tighter but shallower, and that, past a point, turns boxy and one-notey because the trapped air dominates the driver's natural behavior.

This is why the most common complaint about a first closed-back build, thin, boxy, bass-light, is so often a rear chamber that's just too small. The instinct is to reach for foam, but foam won't conjure bass extension that the geometry never allowed. The fix is volume.

How much volume? There's no single number, because it depends on the driver. As a working principle: a driver with a lower free-air resonance and more excursion generally wants a larger rear volume to breathe into, while a small, stiff driver can live in a smaller space. Start from any reference build using a similar driver, then tune from there: your driver and your cup are not theirs.

Putting numbers on it

All of that is intuition, and intuition is where you start. But you still have to pick a number to print, so here are the ballparks I reach for. Treat them as starting brackets for an over-ear closed-back, not targets: they shift with the driver and the design, and the measurement at the end is the only one that counts.

REAR VOLUME	WHAT YOU GET
50-80cc	Tight, controlled bass, resonance relatively high. You'll feel the limit in sub-bass extension.
80-120cc	The sweet spot for most over-ear closed-back builds, deep enough to satisfy, manageable in cup size and weight.
120-160cc	Extended, potentially impressive sub-bass, but cups get large and heavy and resonances get harder to control.

Most commercial over-ear closed-backs land around `~80-130cc`, which is where performance and practicality meet. If you've got nowhere else to start, start there.

The Thiele-Small parameters worth knowing

If your driver has published specs, most headphone suppliers don't publish them, but some Dayton Audio drivers do, and salvaged drivers from good commercial headphones sometimes have documented numbers, a few Thiele-Small parameters give you real predictive power before you print anything.

- `Fs` **(resonant frequency)**: where the driver resonates on its own suspension. Headphone drivers typically land `30-150Hz`. Lower `Fs` means the driver reaches deeper into the bass naturally; a `40Hz` driver extends further than a `120Hz` one, all else equal.
- `Qts` **(total Q factor)**: how damped that resonance is, i.e. how peaked or how broad it is. Above ~`0.7` is a pronounced peak; below ~`0.5` is well-damped and flat. Headphone drivers often sit around ~`0.5-1.0`.
- `Vas` **(equivalent compliance volume)**: the air volume with the same compliance as the driver's suspension. A large `Vas` wants a larger enclosure; a small `Vas` can live in a smaller chamber without going stiff.

A resonance rule of thumb

Here's the one that ties volume and driver together. For a chamber in that `80-120cc` sweet spot, expect the system resonance to land somewhere around `1.0-1.5×` the driver's free-air `Fs`.

So a driver with `Fs` of `60Hz` in a `100cc` chamber rolls off around `75-90Hz`, deep enough for satisfying bass. Drop in a stiffer `120Hz` driver and that same chamber pushes roll-off up toward `150-180Hz`, which is noticeably light in the low end. It's one more reason to weigh `Fs` when you pick the driver, not after.

> TIP
>
> These figures are ballparks. They get you a sensible first print and keep you from guessing in the dark, but the raw measurement off the assembled cup is the final word. If bass rolls off much higher than the rule of thumb predicts, suspect a chamber that's too small or a seal that's leaking.

Sealed or vented

A fully sealed rear chamber gives you the most isolation, but trapping the air also stiffens the spring, and a stiff sealed system tends to throw up a resonance peak, a bump where the bass piles up on one frequency and booms. You can fight that peak three ways: give it a bigger volume, damp it (covered next chapter), or open a vent.

A vent is just a deliberate, controlled leak. Opening a small hole relieves the trapped pressure, lowers and tames that resonance peak, and can smooth the bass into something more even. This is the same sealed-versus-open trade-off from open vs closed back design (Chapter 4), except here you're not choosing one or the other: you're dialing in exactly how much leak you want, by the millimeter. An open-back design is really just the far end of this same slider: a rear chamber so open it barely resists the driver at all.

The catch is that venting trades isolation and bass quantity for evenness. A small vent smooths things; a large one bleeds away low-end punch and lets the outside world in. Where exactly to land that vent is the part you'll fiddle with most.

The front cavity shapes the top end

The front chamber, diaphragm to ear, is set by how deep your cup is and how thick your pad is. It does less for the bass and more for the upper midrange and treble, and for the sense of how close or distant the sound feels. A deeper front cavity tends to move treble peaks downward and push the presentation back, making it feel more spacious or more distant; a shallow cavity brings the sound forward and brighter. This region is also where the resonance that gives headphones their characteristic presence lives, so small changes here are audible.

Because the pad sets so much of this distance, treat pad depth as part of the acoustic design, a sound decision as much as a comfort one: see ear pads and comfort (Chapter 10). Swapping to a thicker pad doesn't just change how the headphone feels on your head; it re-voices the top end.

Design for room to tune

The single most useful principle in chamber design is to build in adjustability, because your first version will be wrong in some specific, fixable way and you want cheap ways to fix it.

You can't easily add volume to a printed cup, but you can always fill or damp to reduce it. So err generous on the rear chamber: print it a little larger than your reference suggests, and bring it down with fill if the bass is loose or the build is too big to wear. It's far easier to take volume away than to add it.

Venting works the opposite way: easy to add, hard to take back once you've drilled it. So design vents you can open incrementally. The cheapest bass-tuning method in this whole hobby is to print the cup sealed but with a few small vent holes plugged, then open them one at a time and listen.

> TIP
>
> Make your rear vents as small drilled holes you can cover with tape, and open them one at a time. It's the cheapest bass tuning you'll ever do: no reprinting, no remodeling, just a roll of tape and your ears. Start fully taped, then peel one vent at a time and listen for where the boom lets go.

Common Mistakes

> CAUTION
>
> - **Rear chamber too small.** The single most common closed-back problem: boxy, peaky, bass-light. Give the driver room to breathe before you reach for damping.
> - **Sealing imperfectly when you wanted sealed.** A closed design relies on a real, continuous seal. A tiny unintended leak collapses exactly the bass you designed the volume to produce.
> - **No venting plan.** Committing to fully sealed with no way to relieve pressure boxes you in if the resonance peak appears. Build in pluggable vents from the start.
> - **Borrowing a volume from another design.** A rear volume that works for someone else's driver won't necessarily work for yours. Use other builds as a starting point, then tune from there.
> - **Treating the front cavity as fixed.** Pad depth is an acoustic parameter. Changing pads re-voices the treble, so account for it when you choose them.

What's Next

The chamber sets the playing field. Damping is how you refine what's happening inside it, taming the leftover resonances and reflections the geometry alone can't address. That's damping strategy and application (Chapter 17), where the chamber you just designed gets its final voicing.

CHAPTER 16

Driver Mounting and Assembly

Mounting the driver, sealing the chambers, and putting the cup together: the hands-on steps where a good design becomes a working headphone.

10 min read   makerphones.com/learn/driver-mounting-and-assembly

You've designed the cup and printed it. Now you put it together. And assembly is where seals are won or lost. The seal is most of the sound, so this stage matters more than its short word count suggests. Rush the seals and a well-designed headphone comes out thin and lifeless, and there's no fixing that in software later. The difference is usually a few minutes of care at a few specific joints.

Cleaning up the print first

Before anything mounts, dry-fit it. A printed driver bore almost never comes off the bed at its exact modeled size: layer lines, elephant's foot at the base, and slight over-extrusion all eat into tolerances. Test that the driver actually seats the way the model intended, and clean up the bore with a hobby knife or light sanding until it does. A driver that's forced into a too-tight bore can sit cocked or stressed; one that rattles in a too-loose bore won't seal. You want it to drop in snugly and sit flat.

Seating the driver

The driver sits in the baffle, and the joint between the driver and the baffle is a seal you cannot afford to leak. Air escaping around the rim of the driver changes the front cavity and bleeds away low end, and because the leak is invisible it's a maddening fault to chase after the fact.

You've got three reasonable ways to mount it. A soft gasket, a ring of foam, rubber, or even a punched craft-foam washer, that the driver presses against as it's held down. An adhesive bead around the rim. Or a mechanical retaining ring or clamp that holds it against a gasket. Lean toward a gasket plus a clamp or screws over gluing, because a glued-in driver is a driver you can't get back out, and you will want to get it back out: to retune, to swap, or to fix a channel you wired backwards.

If your driver has an orientation, some do, with a marked terminal or an asymmetric frame, note it and keep both channels oriented the same way. Consistency between left and right starts here.

Two seals, both critical

Your build has a front seal, driver to baffle, and the cup-to-pad seal against your head, covered in ear pads and comfort (Chapter 10). Both matter, and a tiny leak in either one collapses the low end. When a finished build sounds bass-shy for no obvious reason, a broken seal is the first suspect, every time.

There's a quick field test for it: play steady bass and press the cups gently harder against your head, or against a flat surface. If the bass swells noticeably as you press, you have a leak somewhere in the seal path. Chase it down before you blame the chamber or the driver.

Wiring and polarity

Solder your leads to the driver terminals, and get the polarity right. Tin the leads and the terminals first, keep the iron on the terminal only as long as it takes: driver terminals sit close to delicate parts, and lingering heat is a risk. Run the iron around ~350°C (about 660°F) for lead-free solder, and make each joint quick: too hot for too long risks heat-damaging the voice coil. Mark positive, and keep both channels in phase with each other.

Out at the plug, a standard stereo 3.5mm TRS connector wires up in a fixed pattern. Keep this map handy when you're terminating the cable:

- **Tip**: left positive
- **Ring**: right positive
- **Sleeve**: both grounds/returns (left and right share a common return)

That shared sleeve is the catch with three-conductor TRS: left and right return down the same wire. If your cable is a four-conductor arrangement instead, TRRS or a balanced setup, each channel gets its own separate ground/return rather than sharing the sleeve.

Two channels wired out of phase cancel in the bass and smear the stereo image. The build goes thin and vague, the center of the image dissolves, and it is a inely maddening problem to diagnose if you don't already suspect it: everything measures and looks fine, it just sounds wrong. Checking it takes seconds and saves an evening. Connector and cable choices live in cables, connectors, and hardware (Chapter 12).

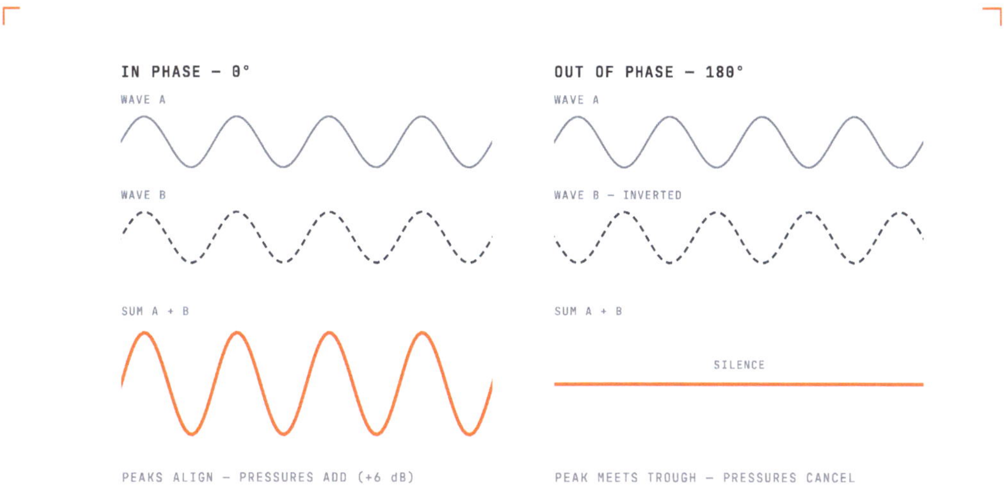

Fig. 16.1 — Why polarity matters: in-phase waves add; out-of-phase waves cancel.

Add strain relief where the wire meets the driver and where the cable enters the cup: a dab of hot glue, a knot, a heat-shrink anchor. The solder joint at the driver is the weakest mechanical point in the whole headphone, and unrelieved cable flex will eventually crack it.

Build the pair identically

Left and right need to match: same mounting, same gasket, same seal, same wiring, same damping. Channel matching is as much mechanical as electrical, and your ears will catch an imbalance you didn't think mattered. The reliable way to get a matched pair is to build both cups the same way, in the same session, with the same materials and the same hands. Don't build one tonight and the other next week from memory.

Order of operations

Damp and wire the driver before you close the cup, and test before final assembly. Start with a multimeter: probe each channel at the plug, and you should read roughly the driver's nominal impedance, ~32Ω for most DIY drivers. An open circuit means a broken or cold joint; a short means you've got a wiring error to chase. Once continuity checks out, wire it up, play both channels with the cup still open, and confirm three things: both sides make sound, polarity is correct (a mono track should image dead center), and there are no rattles or buzzes from a loose part or a diaphragm fouling on something. Only then commit to sealing or gluing anything shut.

> TIP
>
> Before you glue or close anything, wire it up and play both channels with a mono track. It should image as a single point right between your ears. If it sounds wide, hollow, or sucked-out, one channel is out of phase. Fix it now, while the cup is still open, not after you've glued it shut.

Closing up

If your cup is a two-part print, the seam between the halves is another seal, and so is the joint where the baffle meets the cup body and where the cable enters. Whether you're gluing, screwing, or press-fitting, make those joints airtight on a closed design: a bead of glue, a gasket, or a wrap of tape on the cable pass-through. On an open design you have more latitude back there, since the rear is meant to breathe, but the front seal and channel matching still have to be right.

Common Mistakes

> ⚠ CAUTION
>
> - **Reversed polarity on one channel.** Phase cancellation: thin sound, gutted bass, vague imaging. Confirm both channels are in phase with a mono track before closing up.
> - **Gluing the driver in permanently.** You'll want to retune or swap it. Use a gasket and clamp so the build stays serviceable.
> - **Ignoring the front seal.** A leak between driver and baffle changes the sound as much as a bad chamber does. Seal it and test for leaks.
> - **Mismatched cups.** Different gaskets, damping, or seal between left and right gives you a channel imbalance you'll hear but struggle to place. Build them as a matched pair, together.
> - **No strain relief.** Cable flex cracks the solder joint at the driver, the most common long-term failure. Anchor the wire at both ends.

What's Next

With the driver mounted, sealed, and tested, the chamber and damping decide the final voice. Damping strategy and application `(Chapter 17)` is where you take a sealed, working build and dial in how it actually sounds.

CHAPTER 17

Damping Strategy and Application

Where and how to apply damping in your build to fix specific problems: a strategy for using felt, foam, and fill instead of just stuffing the cup.

10 min read makerphones.com/learn/damping-strategy-and-application

Damping is the finest tuning lever you have, and the easiest one to overdo. Damping materials (Chapter 11) covered what to use: felt, open-cell foam, fiber fill, and what each does to sound. This chapter is about where to put it and why, which is a different and more useful skill than knowing the materials. A builder who understands placement can tune a headphone with a single sheet of felt; a builder who doesn't will stuff a cup full of foam and wonder why it went dead.

Damping solves problems, not "general badness"

The mistake almost everyone makes is treating damping as a way to make a build "better" in some vague, undirected way: add a little foam, see if it improves, add a little more. That's the road to an over-damped, lifeless headphone, and you arrive there one reasonable-seeming step at a time.

Instead, name the specific problem first. Is there a peak that makes one note boom or one region shout? Ringing that smears detail and makes the sound hard? Too much treble bite? A boxy, congested midrange? Each of those has a placement that addresses it. And if you can't name the problem, you're not ready to add material yet.

How to tell what you're hearing

A few rough signatures help you point the damping in the right direction. A resonance peak tends to sound like one frequency jumping out louder than its neighbors: a boom on a particular bass note, or a hard edge on a particular vowel. Ringing sounds like a smear or a metallic hardness, energy that hangs on a beat too long and blurs fine detail. Boxiness is a cupped, hollow, talking-into-a-jar coloration in the lower midrange. Excess treble is obvious enough: it bites, sibilance spits, cymbals turn to hiss. None of this is precise, but it's enough to choose a starting placement, and once you bring measurement in next section you'll see these problems instead of only hearing them.

Where it goes and what it does

Work from coarse to fine: the placements with the biggest effect first, the delicate ones last.

Behind the driver, in the rear chamber. This is the highest-impact placement. Fill here absorbs the rear wave, tames the main resonance, and kills boxiness. More fill means more control, but past a point it strangles the bass and drains the life out of the sound, because you're absorbing energy the driver needs. This is your main lever and the one to approach most carefully.

On the cup walls. A lining of felt or thin foam kills reflections and wall resonances, cleaning up the midrange and reducing hardness. This is usually safe and effective, and it's the right place to start if the build sounds congested or hard rather than specifically peaky.

Over a vent (resistive damping). Covering a vent or port with a layer of felt or foam doesn't block it: it adds resistance, slowing the airflow through it. This is how you fine-tune a vent's effect without changing its size, and it's a gentler control than drilling.

In the front cavity, around the driver. This affects treble, and it's the riskiest spot. A thin ring of felt can tame a sharp treble peak, but a little too much and the whole top end goes dull and closed-in. Use it sparingly, deliberately, and never as a default first move.

Less is more

Add damping in small increments. Make one change, then listen or measure, then decide on the next. The target is the *smallest* amount of damping that solves the problem, not the most you can fit in the cup.

Over-damping gives you a dead, airless sound with no sparkle and no sense of space. Under-damping leaves you ringy, peaky, and boxy. The sweet spot between them is narrower than beginners expect, and you find it by sneaking up on it from the lean side, not by stuffing the cup and backing off, because once a build is over-damped, every further change feels like it isn't enough, and you keep adding.

Keep it reversible while you tune

Use friction-fit felt and loose fill while you're dialing things in, so you can pull material back out as easily as you put it in. Cut pieces a little oversized so they hold by friction, and keep notes on what's in each cup. Don't glue anything down until you're confident the build is where you want it. Once it's glued, you've locked in a guess. Keep it removable and you can change your mind for free.

> **TIP**
>
> Change one thing at a time. Add felt to the rear chamber, listen, then decide on the next move. If you change three things at once you'll never know which one mattered, and you'll almost always end up over-damped, because every individual change felt like "a little more couldn't hurt."

What damping can't do

Damping refines a chamber; it can't rescue a badly wrong one. If the rear volume is too small or the seal is broken, no amount of felt fixes it. Go back to acoustic chamber design `(Chapter 15)` or check the seal first. Builders waste enormous effort trying to damp their way out of a geometry problem. Reach for damping once the geometry is right, as the final voicing step, not as a patch for a chamber that was never going to work.

Common Mistakes

> ⚠ CAUTION
>
> - **Stuffing the cup full "to be safe."** This kills bass and life. More damping isn't safer – it just trades one problem for a worse one (dead bass).
> - **Asymmetric damping between channels.** Different fill in left and right gives you an imbalance you'll hear but can't easily place. Match them exactly.
> - **Using damping to fix a wrong chamber.** It can't. Sort the volume and the seal first, then voice with damping.
> - **Gluing before tuning.** Commit only once you're sure. Keep everything removable while you work.
> - **Chasing a treble peak into a dull top end.** A thin ring is plenty up front. Too much felt and you've traded a peak for a lifeless treble, a worse trade than you think.

What's Next

You can do all of this by ear, and many great builds are tuned exactly that way. But the moment you want to see what each change actually does, and stop guessing, you bring in measurement. Tuning with damping (Chapter 22) is the measurement-driven version of everything here: measure, change one thing, measure again, and watch the problem shrink.

PART

4

Measurement & Tuning

CHAPTER 18

Why Measure Headphones

What measurement gives you that listening can't, and what it can't give you, plus why even a rough rig beats guessing once you're tuning a build.

10 min read   makerphones.com/learn/why-measure-headphones

Measurement means running test tones close to your ears. Keep Listening Safely in mind before your first sweep.

You can build and tune a good headphone entirely by ear, and plenty of great builds happen exactly that way. But there's a point in every build where measurement stops being optional and starts saving you hours: the point where you need to know *what's* actually happening. Measurement turns a vague feeling into something you can see and point at. That's why you pick up a mic.

This chapter is about why measuring is worth the trouble, what it can and can't tell you, and how to hold your own measurements at the right level of trust, because a rig you misread is worse than no rig at all.

Two different questions

Your ears and a measurement rig don't answer the same question, and a builder needs both answers. Your ears tell you whether you *like* the sound: that's the only verdict that ultimately matters, and no graph overrides it. A measurement tells you *what the headphone is actually doing*: where the energy sits, where the imbalances are, what changed when you added felt. One's about taste; the other's about what's going on. Lean only on your ears and you'll guess wrong about why something sounds off. Lean only on the graph and you'll tune toward a number that sounds like hell.

What measurement gives you

Channel matching. Ears are surprisingly bad at catching small left-right imbalances. You adjust to them within seconds and stop noticing. A measurement shows a mismatch instantly and unambiguously. Matched channels are the difference between a stable, centered image and one that drifts or feels subtly lopsided, and matching is one of the things measurement does that listening can't do reliably.

Seeing the problem. "The bass sounds boomy" is a feeling. A measurement shows you the peak, which frequency, how tall, how wide, so you can target it directly instead of guessing at fixes. This is the single biggest day-to-day reason to measure while tuning, and it turns a frustrating evening of trial and error into a few deliberate moves.

Comparing to a target. You can overlay your curve against a target, a research-based preference curve, a commercial headphone you like, or your own reference, and see precisely where and by how much you deviate. Understanding frequency response (Chapter 2) covers what those curves mean and why flat is not the goal.

Documenting your work. Measure before and after each change and you build a record of what every tweak actually did. That record is what lets you back out of a change that made things worse, compare two builds honestly, and avoid repeating a mistake three sessions later. Memory is unreliable about sound; a saved measurement is not.

What measurement can't do

A measurement can't tell you whether you'll enjoy the headphone. A technically clean curve can sound wrong to you, and an unusual one can sound wonderful: preference is real and personal, and the graph has nothing to say about it. The curve helps you understand what's happening. It's not a score to chase.

And your rig is not an industry-standard measurement head. A hobby setup gives you useful, repeatable *relative* information, but its absolute numbers are off, and some regions are barely trustworthy at all. The high treble especially, roughly the top octave, is extremely sensitive to your specific coupler, the exact seating, and tiny positional changes, so wild wiggles up there are often artifacts of your setup rather than real features of the headphone. Read your measurements as a reliable guide to *changes* and *shapes*, lean on them hard for that, and always confirm the final result with your ears.

You're not chasing flat

It's worth saying plainly, because the graph constantly tempts you toward it: the goal is not a flat line. We don't hear flat as neutral. The ear canal and the shape of the outer ear impose their own large response, a broad boost in the presence region and a roll-off pattern up top, which is exactly why measured target curves are tilted and bumped the way they are rather than ruler-straight. You're aiming at a target that accounts for how human hearing works, not at a horizon line. Tuning a headphone until its raw curve looks flat is a reliable way to make it sound thin and harsh. The listening research that produced those targets is mapped in Sources & Further Reading.

Calibrate your expectations

The most useful first thing to do with a new rig isn't to measure your build: it's to measure a commercial headphone you know well. Seeing how a headphone you already understand looks on *your* rig teaches you what your particular setup does: where it reads high, where it reads low, how much the top octave jumps around between seatings. After that, every measurement of your own builds means more, because you know how to read your own instrument.

When you actually need it

You don't need a rig for your first listen-and-tweak, and you shouldn't let the lack of one stop you from starting. But the moment you're seriously voicing a build, matching channels, or comparing one build against another, measurement goes from nice-to-have to indispensable, and most builders reach that moment faster than they expect.

> TIP
>
> Measure before *and* after every change, and save each file with a name that says what you did. The overlay of "before versus after" is the most useful single thing measurement gives a builder. It shows you exactly what your tweak changed, with no guesswork and no reliance on memory, which is the part of tuning your ears are worst at.

Common Mistakes

> ⚠ CAUTION
>
> - **Trusting absolute numbers from a hobby rig.** Read changes and shapes, not exact dB values as though they came from a reference head.
> - **Tuning to flat.** Chasing a ruler-straight graph fights the way you actually hear. Aim at a target curve instead.
> - **Not saving measurements.** Without a record you can't compare, can't back out a bad change, and will repeat your own mistakes.
> - **Believing a single measurement.** Positioning varies seating to seating. Re-seat and remeasure before you act on a small feature. It may not be there the second time.
> - **Skipping the known-headphone check.** Measure something familiar first so you know how to read your own rig before you trust it on a build.

What's Next

Knowing why to measure is one thing; getting a rig you can trust without spending thousands is the next, and a budget measurement setup, what to buy and how to assemble it, is where this section goes from here. Once you can measure, tuning with damping `(Chapter 22)` puts the rig to work on an actual build.

CHAPTER 19

Budget Measurement Setup

Building a headphone measurement rig on a maker's budget: couplers, mics, software, and the parts you can 3D-print yourself, without the five-figure industry gear.

11 min read makerphones.com/learn/budget-measurement-setup

The previous chapter made the case for measuring. This one is about what you actually put on the bench to do it, on a DIY budget. The industry rigs run from roughly ten thousand dollars into the tens of thousands. The good news is that a usable headphone measurement setup costs a few hundred dollars or much less, and because you're a maker, you can 3D-print a real part of it yourself.

What a setup is made of

Every measurement rig, cheap or expensive, is the same chain. Something drives the headphone, the headphone plays into an artificial ear, a microphone in that artificial ear captures the sound, and a computer running measurement software turns it into a graph.

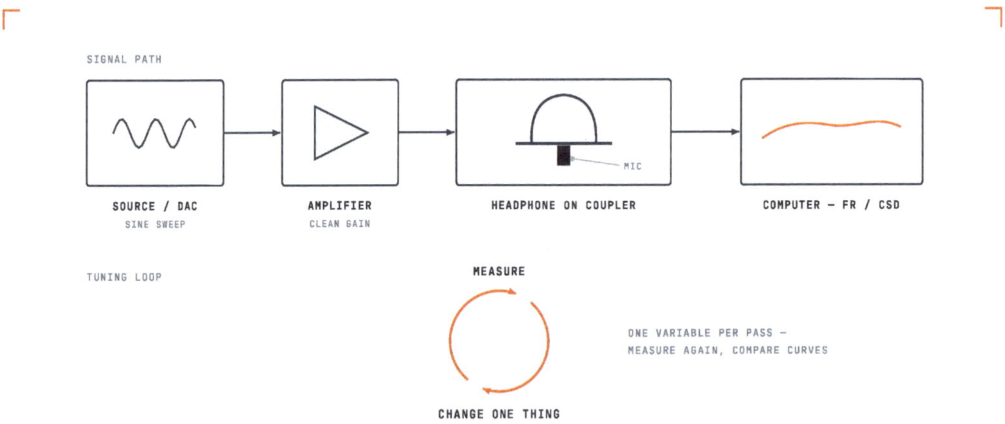

Fig. 19.1 – The measurement chain every rig shares, and the loop that does the tuning.

On the playback side you need a way to drive the headphone: any decent small DAC or headphone amp. One modern note worth absorbing: use an amp with very low output impedance. The old advice to measure through a 120-ohm output is obsolete; today you want the lowest output impedance you can get, so the amp isn't coloring the result.

On the capture side you need a coupler, an artificial ear canal with a microphone, and, for over-ear headphones, a silicone pinna (an artificial outer ear) for the headphone to seal against, plus a stand to hold everything in alignment.

The software is the easy part: Room EQ Wizard, universally called REW, is free and the de facto standard. Everything below works with it.

The pieces

The coupler and mic. This is the heart of the rig. The affordable-but-serious choice now is an IEC 711 clone coupler: an unofficial coupler built to the IEC 60318-4 standard, widely available online for around a hundred dollars, that either includes a microphone or pairs with one. This is what most of the hobby measures with today, and it's the path to data you can actually trust within its limits.

The pinna, for over-ears. A silicone artificial ear that the earpad seals against, sitting in front of the coupler. IEMs don't need it — they couple directly. But over-ear headphones do, because the pinna and the seal are part of what you're measuring.

The stand. Something to hold the coupler, pinna, and headphone in consistent alignment. Here's the maker's advantage: you can print this. The community has shared fully 3D-printable measurement stands designed around the common clone couplers and pinnae, so the fixture that would otherwise be an awkward purchase becomes an afternoon on your own printer, and one you can tweak to fit your exact coupler.

Getting the signal into the computer. The coupler's microphone reaches your computer one of two ways: through a USB measurement microphone that's class-compliant and shows up as an input on its own, or through a separate USB sound card if what you have is a bare electret capsule. Bare electrets need plug-in power, a small bias voltage, which a sound card's microphone input or a little audio interface supplies. Either way, the goal is a clean, calibrated input that REW can see, with the recording level set so a measurement sweep sits comfortably above the noise floor and well below clipping. Set the level once and you won't have to touch it again.

Three budget tiers

Shoestring, to learn the ropes (a few dollars to a few tens). A small calibrated electret measurement microphone plus a coupler you build yourself from vinyl and latex tubing. You tune the tube length so the canal resonance lands somewhere around 8 kHz, apply the mic's calibration, and you're measuring. The limits are real: the cheap mic rolls off in the deep bass, and it conforms to no standard, but it is genuinely good enough to see your own before-and-after changes and to match channels, which is most of what tuning needs early on. As a way to learn the workflow before spending, it's hard to beat.

The sweet spot (a few hundred dollars). An IEC 711 clone coupler, a silicone pinna for over-ears, a 3D-printed stand, REW, and a low-impedance amp to drive the headphone. This is what serious DIY builders use, and it's the target to aim for once you know you'll keep at it.

The turnkey all-in-one (from around $180 to well into four figures). Two very different products live here. The original miniDSP EARS, around $180, is a plug-and-play USB jig that needs no assembly, but be clear-eyed: it predates the clone-coupler era, it isn't built to any international standard, and it's fine for fun and relative work, not for anything you want to trust deeply or share. Its successor, the miniDSP EARS Pro, is a genuinely different animal: a full upright fixture with an adjustable stand that *is* built to the IEC 60318-4 standard, which makes it more reliable and consistent than the AliExpress clones, with no assembly and a precision calibrated capsule. It can also be calibrated against an SPL reference to read absolute levels, actual sensitivity, and distortion at a known volume, which the cheaper rigs can't manage. The catch is price: the fixture is about $1,550, or roughly $1,800 with the USB audio interface you'll want. And since it ships from China, expect an import tariff added at checkout, recently around $450 for US orders, which brings it closer to $2,000–2,250 delivered. That's far above the budget tiers, though still a fraction of the $10,000-plus industry rigs it sits just below in accuracy. For most builders on a budget the clone coupler or the shoestring DIY rig is still the answer; the EARS Pro is for people who'd rather pay for a finished rig than build one. And, fittingly for this crowd, you can also put together an equivalent IEC 60318-4 rig yourself from a coupler, pinna, mic preamp, and a 3D-printed stand.

Calibrate, and stay humble about it

Clone couplers and budget mics vary from unit to unit in manufacturing, so you apply the supplied calibration file and treat your rig as *your* reference, not a universal one. This connects directly to the honesty from the last chapter, now made concrete: even a good clone rig is reliable only up to roughly 10 kHz, and the extremes, the top treble especially, and to a degree the deep bass, should be read with caution, because peaks and dips up there can land at the wrong frequency or the wrong height.

The practical consequence: rig-to-rig differences mean you can't directly compare your graphs against a reviewer's or another builder's. Your measurements are for tracking *your* changes and matching *your* channels.

> TIP
>
> Start at the shoestring tier, an electret mic and a tubing coupler, before you spend on a clone coupler. It's enough to see your before-and-after changes and match channels, which is the large majority of what tuning actually needs, and it teaches you the whole workflow cheaply before you put money into the better rig. You'll buy the clone coupler with much clearer eyes for having done it.

Common Mistakes

> ⚠ **CAUTION**
>
> - **Treating hobby-rig numbers as industry-standard.** They're useful and relative, not absolute. Read changes and shapes, not exact values.
> - **Comparing your graphs to reviewers'.** Different rigs aren't comparable. Use your measurements to track your own builds.
> - **Skipping the calibration file.** The clone coupler and mic need their correction applied, or the result is wrong from the start.
> - **Trusting data above ~10 kHz.** That's past where these rigs are reliable. Be skeptical of fine detail and exact peak positions at the extremes.
> - **Measuring through a high-impedance amp.** The old 120-ohm advice is outdated. Use the lowest output impedance you can.

What's Next

A rig is only as good as your technique with it. The same headphone measures differently depending on how you seat it, and a graph only helps once you know how to read it, so the next chapter is about taking clean, repeatable measurements and actually interpreting them: taking and interpreting measurements (Chapter 20).

CHAPTER 20

Taking and Interpreting Measurements

How to take headphone measurements you can actually trust: seating, repeatability, smoothing, and how to read the resulting graph instead of chasing noise.

11 min read makerphones.com/learn/taking-and-interpreting-measurements

A measurement rig is only as good as your technique with it. The same headphone, on the same rig, will give you noticeably different graphs depending on how you seat it. And a graph you can't read is just a wiggly line. This chapter is the bridge between owning a rig and getting something useful out of it: how to take measurements you can trust, and how to interpret what you see.

Set up once, correctly

Before any of it means anything, get the basics right. Load your microphone and coupler calibration file in REW so the rig's known errors are corrected. Set the SPL calibration so your levels are meaningful rather than arbitrary. Drive the headphone from a low-impedance amp. And use a logarithmic sweep, REW's standard measurement, at a sensible level, loud enough to sit well above the noise floor but nowhere near clipping. Once this is dialed in, it stays put, and you can stop thinking about it.

How you seat it changes the graph

Here is the single biggest source of variation in headphone measurement: how the headphone sits on the rig. Clamping force, centering, and the quality of the seal against the pinna all move the result, and they move it most in the regions that matter. So the trick is doing it the same way every single time. Place the headphone the same way, with the same downforce and the same seal.

And then don't trust a single placement. Take several measurements, lifting and re-seating the headphone between each one. The headphone's real features show up in the same place every time. The stuff that jumps around between seatings is the fit and the rig, usually up in the treble, which is exactly where these rigs are least reliable anyway. Overlay your seatings, or average them, and let the agreement between them tell you what's real. Where three measurements land on top of each other, you're looking at the headphone. Where they scatter, you're looking at the seating.

Measure both channels under identical conditions, too, same seating routine, same level, or any left-right comparison you make later just measures your own technique and won't tell you anything about the build.

Smooth it, or you'll chase noise

A raw measurement is far too spiky to interpret: every tiny ripple, real or not, is on display. Apply smoothing, somewhere in the range of a twelfth to a twenty-fourth of an octave, or REW's psychoacoustic smoothing, so that what you see is the shape rather than the hash. You're trying to read the headphone's character, and the character lives in the broad strokes, not in needle-thin spikes that no ear would ever resolve.

Reading the graph

Work from the big picture down. First take in the overall tilt, the broad balance from bass to treble, because that, more than any single feature, is what you hear as the headphone's signature. Then look at the major features: the bass shelf, any obvious peaks or dips, the presence region up in the upper midrange and lower treble.

Then compare to a target. Overlay a target curve, a research-based preference curve (see Sources & Further Reading for where these come from) or your own reference, and read your measurement as *deviation from that target,* not as distance from a flat line. As covered earlier, flat is not the goal, and a graph read against the right target tells you far more than one read against zero. Where you sit above the target, you have too much; where you sit below, too little.

Overlay left against right as well. The two channels should track closely; where they diverge you have a matching problem to chase back to the build: a seal, a damping difference, a driver mismatch.

Throughout, keep the trust calibrated. Believe the broad shape and, above all, believe your before-and-after deltas. Those are what your rig is genuinely good at. Be skeptical of fine detail above roughly 10 kHz and of the exact frequency and height of features at the extremes, where the rig's limits live.

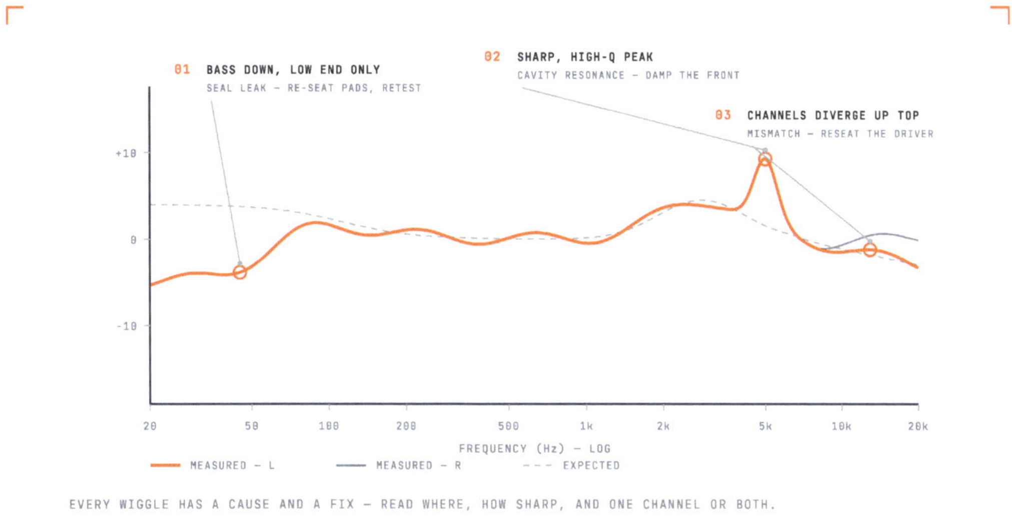

Fig. 20.1 – Reading a measurement as diagnosis, three flagged features (illustrative trace), each pointing at a cause and a fix: a seal leak, a cavity resonance, a channel mismatch.

Turning features into fixes

A graph is only useful if it points at an action, and by now you have the chapters to act. A sharp peak is a resonance. Head for resonance control (Chapter 27) and your damping. A boomy, oversized bass shelf points back at the chamber and venting. A left-right divergence is an assembly and seal problem. And the whole measure-read-adjust loop comes together in tuning with damping (Chapter 22), where each measurement drives one targeted change and the next measurement checks it.

Keep a record

Save every measurement with a name that says what it was: the build, the date, the change you'd just made. REW lets you keep many traces open at once, so you can overlay tonight's result against last week's and see the whole arc of a build rather than a single snapshot. A folder of well-named measurements is what turns a pile of graphs into a history you can actually learn from, and it's the same habit that makes the tuning loop work and repeatable builds possible. The five seconds it takes to name a file is nothing against the hour it saves when you're trying to remember what you did three sessions ago.

When a measurement is good enough

You don't need perfection, and over-measuring wastes evenings. A measurement is good enough when it's consistent across re-seatings, the channels track, and the shape makes physical sense for what you built. When those three hold, stop measuring and start acting, or, if the build is where you want it, stop and go listen.

> TIP
>
> Take at least three measurements, re-seating the headphone each time, and overlay them before you trust anything. That one habit catches more bad data than any gear upgrade ever will.

Common Mistakes

> ⚠ CAUTION
>
> - **Trusting a single seating.** One placement isn't data. Re-seat several times and overlay to see what's stable.
> - **Reading un-smoothed graphs.** Raw curves are all noise. Smooth them so you read shape, not hash.
> - **Comparing to flat instead of a target.** Read deviation from a target curve, not distance from zero. Flat isn't the goal.
> - **Over-interpreting the top octave.** Above ~10 kHz the rig is unreliable. Don't chase peaks and dips up there.
> - **Mismatched left-right conditions.** Measure both channels the same way, or your matching comparison is meaningless.

What's Next

You can now take a measurement you trust and read what it's telling you, which is exactly what the tuning loop runs on. Tuning with damping (Chapter 22) puts this to work: measure, make one change, measure again, and watch the build move toward your target.

CHAPTER 21

Measuring a Raw Driver on the Bench

Build a simple test baffle and use REW to measure raw headphone drivers, frequency response, impedance, and distortion, before you build.

9 min read   makerphones.com/learn/measuring-raw-drivers

Before a driver ever goes into a cup, you can learn a surprising amount about it on the bench. A few quick measurements tell you whether a driver is healthy, how two samples compare, and whether a candidate is even in the ballpark for what you're building. Spend ten minutes here and you'll stop wasting whole builds on a driver that was never going to work.

What a bench measurement actually tells you

A raw driver measured in open air characterizes the driver; it doesn't predict the sound. It won't tell you how the driver will sound on a head: that comes from the cup, the pads, the leak around your ears, and the small trapped volume at the eardrum, none of which exist on the bench. The bass especially is a product of that coupling, so don't read much into the low end of a free-air curve.

What it does tell you is plenty: the general shape of the midrange and treble, where the main resonance sits, where the cone or dome starts to break up, how sensitive the driver is, and, maybe most useful, how consistent two samples are. That's your baseline. Everything you do later is measured relative to it.

Think of it as triage. You're not voicing anything yet. You're answering one question: is this driver healthy, and is it in range?

A simple test baffle you can build

The rig is a baffle: a flat panel with the driver firing through it and the back open to free air. The one I use is two layers of half-inch Baltic birch glued into a roughly one-inch slab, about a foot square, raised up on legs.

It's thick and laminated for a reason. A thin panel rings, and that ringing lands right in your measurement. Mass and stiffness keep the baffle quiet so the only thing you're measuring is the driver.

The center has a removable plate, around five inches across, that the driver mounts to from underneath. That drop-in plate is the whole trick: you can swap drivers in seconds and keep everything else identical. Mount the driver airtight to the plate, and seal the plate to the baffle. Any leak moves the low end around and quietly ruins a comparison.

Round over or chamfer the edges, especially where the plate meets the panel. A hard ninety-degree edge causes diffraction that ripples your response; a forty-five-degree bevel or a generous roundover tames it. Then raise the whole thing on legs, out in the open and away from walls. The higher and clearer it sits, the longer you get before the first reflection arrives, which matters for the next step.

Don't bother going bigger. A larger baffle buys you trustworthy response a little lower in frequency, but that's not the range that sorts good drivers from bad, and free-air bass doesn't predict headphone bass anyway. If you ever want a clean low end, measure nearfield, mic a few millimeters off the dome, instead of building a bigger panel.

The measurement chain

You need three things: the baffle, a measurement mic, and REW (Room EQ Wizard, free). A USB measurement mic keeps it simple: plug it in, no audio interface required.

Calibrate the mic first. This is the step people skip and then chase problems that were never real. Your mic shipped with a calibration file matched to its serial number; load it into REW's mic settings before you measure anything. Without it, the top octave will lie to you.

Then lock your geometry down and don't touch it. Pick a mic distance and an on-axis spot over the driver and keep it identical for every driver you test. For comparison work, repeatability beats absolute accuracy every single time.

Set your levels so REW's input peaks land around −12 dBFS, and drive every driver at the same level, otherwise the sensitivity numbers aren't comparable. Run a log sweep from 20 Hz to 20 kHz.

Last, gate the measurement. Open the impulse response REW captured and set the window to end just before the first reflection shows up. That window sets the lowest frequency your gated curve is honest down to, and the higher your baffle sits, the lower that limit goes.

Two measurements worth adding

Frequency response is only part of the story. Two more cheap measurements do most of the actual screening work.

Impedance is the big one. An impedance sweep is repeatable and doesn't care about your room or your mic. It hands you the free-air resonance (Fs), the voice-coil DC resistance, and the height and sharpness of the impedance peak. A driver with a rubbing or off-center coil gives itself away immediately as a ragged or doubled peak. The easy way to measure it is a Dayton DATS: a little USB pod you clip the driver leads to; it runs a sweep and reads out the numbers. You can also do it in REW with an interface and a sense resistor, but that's more fiddling and you have to mind your levels.

Distortion is the other. REW reports harmonic distortion straight from the same sweep you already ran. A driver with a pinched surround, a rub, or a rough batch lights up with distortion a clean sample doesn't. That's your dud filter, in numbers.

Putting it to use

Overlay everything in REW. Stack several samples of the same driver on one graph and watch how tightly they track through the midrange and treble. For a left/right pair, pick two that sit within a decibel or so of each other, mismatched channels are something you'll hear, and something you can't easily fix later.

Then compare different candidates against each other and against whatever you were hoping for. You're not looking for perfection; you're looking for healthy, consistent, and roughly in range. A driver with a wild peak, a ragged impedance curve, or high distortion is telling you something. Believe it, and set it aside.

That's the entire point of the bench: ten minutes here means you'll never again find out a driver was a dud *after* you've built a headphone around it.

Common Mistakes to Avoid

Mounting the driver loose. A leaky or rattling mount shifts the low end and adds noise. Seal it and snug it, every time.

Skipping the mic calibration file. Your treble will look wrong and you'll burn an afternoon trying to fix a driver that was fine all along.

Expecting real bass from a small baffle. Below a few hundred hertz the panel stops loading the driver properly. That roll-off is the fixture talking, not the driver.

Comparing measurements taken differently. Different mic distance, different drive level, different setup, and the comparison means nothing. Lock it down and leave it.

Treating a free-air curve as the final sound. It's a baseline. The cup, the pads, and your ear reshape all of it, the bass most of all.

What's Next

Once you can trust a driver, the next step is measuring it loaded, in a cup, with pads, the way it'll actually be heard. That's where a coupler or chamber comes in, and it's covered in acoustic chamber design `(Chapter 15)`.

If you're still assembling your bench, the budget measurement setup `(Chapter 19)` walks through the gear, and taking and interpreting measurements `(Chapter 20)` goes deeper on reading the curves. And if you're not yet sure which driver to put on the baffle in the first place, start with the driver selection guide `(Chapter 7)`.

Difficulty: Intermediate **Prerequisites:** a working handle on frequency response `(Chapter 2)` and impedance and sensitivity `(Chapter 3)`

Related articles:

- Why measure headphones `(Chapter 18)`
- Budget measurement setup `(Chapter 19)`
- Taking and interpreting measurements `(Chapter 20)`
- Acoustic chamber design `(Chapter 15)`
- Driver selection guide `(Chapter 7)`

Put a driver on the bench

Build the baffle, load your calibration file into REW, and measure the first driver in your parts bin. Then measure a second one and overlay them. That single graph is the start of everything that follows.

CHAPTER 22

Tuning with Damping

The measurement-driven tuning loop: read the curve, make one targeted change, remeasure, repeat. How to turn a measurement into a fix, and when to stop.

10 min read makerphones.com/learn/tuning-with-damping

With a rig you can trust and some damping material on the bench, you can tune by data instead of by guess. The loop itself is simple, almost embarrassingly so. The discipline to run it cleanly is the entire game, and it's where most builders trip, by changing too much at once and learning nothing from any of it.

Everything in damping strategy and application (Chapter 17) about where damping goes and what it does still applies here. This chapter just adds a measurement at every step, so you're acting on what you can see rather than what you hope is true.

The loop

Measure. Read the curve. Make one targeted change. Remeasure. Compare. Repeat.

That's the whole method. The skill that makes it work is changing a single variable at a time, so that when the curve moves, you know exactly what moved it. Change three things between measurements and you're back to the same guessing game you were trying to escape, only now there's a graph to make you feel productive.

Measure honestly before you start

A measurement-driven loop is only as good as the measurement. Before you trust any curve, take it a few times, re-seating the headphone between each, and see what stays put and what jumps around. The features that survive re-seating are real and worth chasing. The ones that move are positional artifacts, usually up in the top octave, and chasing them is how you waste an evening tuning something that was never there. Establish your repeatable baseline first, then start changing things.

Reading the curve and acting on it

A peak. A sharp bump usually means a resonance. Address it with damping behind the driver or on the wall nearest it, or with venting if it sits low. Add material in small amounts and watch the peak shrink. When it stops shrinking for the felt you're adding, you've found the limit of that placement.

Boomy or excessive bass. Often a chamber resonance or just too much low-end energy. Try opening or adding resistance to a vent, reducing rear fill, or adding targeted damping. The bass region responds strongly to small changes, so move in small steps and remeasure often.

Ringing. Energy that hangs on after the signal stops. You'd see it most clearly on a decay plot, which the advanced measurement chapter covers. Wall lining and rear fill are your tools, and the payoff is a cleaner, less hard sound even when the frequency response barely moves.

A treble peak. A thin damping ring in the front cavity, applied sparingly. Treble is where over-damping does the most audible damage, so make small changes and check both the curve and your ears after each one.

Channel mismatch. Overlay the left and right curves and match the damping and seal until they sit on top of each other. This is pure mechanical matching, same material, same placement, both sides, and the graph makes a mismatch you'd never hear plainly visible.

Reading the problem, choosing the material

The advice above tells you *which way* to move; this table tells you *what to reach for*. Different frequency problems respond to different materials in different places, and reaching for the wrong one wastes an evening. Match what you measure to the first thing worth trying, then run the loop.

WHAT YOU MEASURE	WHAT TO TRY FIRST

Boomy / elevated bass (80–300Hz)	Foam on the rear chamber walls: start with one wall. An alternative for excess mid-bass is a small amount of cotton fill in the rear chamber, acting as an acoustic mass that smooths the resonance.
Hollow / boxy midrange (300Hz–1kHz)	Chamber standing waves. Thin felt on the interior side walls often helps more than foam here: felt absorbs well in the mids without killing everything else. Try one wall at a time.
Forward / intense upper mids (1–4kHz)	Swap to pads with a more absorptive inner surface (velour over pleather), or add padding on the front baffle around, not over, the driver to cut reflections.
Harsh / sibilant treble (4–8kHz)	Thin felt in front of the driver, applied sparingly. Start small: a 2cm circle of craft felt across part of the driver face is enough to see an effect. This spot is sensitive, so go slowly.
Narrow resonance spike (anywhere)	Find the source first. Tap the cup with the driver out and listen for a tonal ring. That's the mode. Treat it at the chamber, or in front of the driver if the break-up is in the driver itself.

> ⚠ **CAUTION**
>
> A *dip* is usually not a damping problem. Damping removes energy; it can't add it, so no amount of foam will fill a hole in the response. A thin low end means the headphone isn't producing enough down there: check the seal, the chamber volume, and the pad depth, or change the driver. That's a design fix, and no amount of damping substitutes for it.

What one pass actually looks like

Say your build measures with a tall, narrow peak in the upper bass and an otherwise reasonable curve. You take it three times, re-seating between each, and the peak holds steady every time, so it's real, not an artifact. You add a small amount of fill behind the driver, nothing else, and remeasure with the same seating routine. The peak drops by a few decibels and widens slightly; the rest of the curve barely moves. Good. That's the placement working. You note the change in your log, listen to confirm the boom has eased without the bass going lifeless, and decide whether the remaining peak is worth another small pass or whether you've reached the point of diminishing returns.

String a handful of those clean passes together and you've tuned a headphone deliberately instead of by luck. The temptation is always to skip the remeasure, or to add felt in two places at once "to save time," and both shortcuts cost you far more time than they save, because they take away your ability to know what worked.

One change at a time

This is the rule that makes the whole method work, so it bears repeating: change one thing, then measure. Overlay the new curve on the previous one and look only at the difference. If you change three things between measurements, you've learned nothing reliable about any of them, and you'll drift toward over-damping, because each individual change felt harmless on its own.

Knowing when to stop

Stop when the curve is close to your target *and* it sounds right to you. Both conditions have to be met together; either one alone will mislead you. Chasing the last decibel on the graph past the point where it sounds good is exactly how you over-damp a build into lifelessness: the graph improves while the music gets worse.

The first few changes do most of the work, and after that you hit diminishing returns fast. Obsessing over tiny wiggles isn't worth it, especially since many of those wiggles are your rig's artifacts rather than anything a listener would ever notice. Knowing when a build is *done* is as much a skill as knowing what to change, and it's mostly the discipline to stop when both your eyes and your ears agree.

> **TIP**
>
> Keep a log. Name each measurement file with what you changed: "rear-fill-2g," "wall-felt," "front-ring-thin." Three sessions in, that log is how you remember what worked, avoid repeating a change that didn't, and stop going in circles. It's the cheapest tool in the whole process and the one builders most often skip, right up until the night they wish they hadn't.

Common Mistakes

> ⚠ CAUTION
>
> - **Changing multiple things between measurements.** The cardinal sin. You can't attribute the result to any one change. One variable at a time, every time.
> - **Tuning to the graph until it's flat.** If it measures great and sounds dull, the graph won the argument it shouldn't have. Your ears make the final call.
> - **Chasing artifacts.** Before you act on a tiny feature, re-seat the headphone and remeasure. If it moves, it wasn't real. Leave it alone.
> - **Over-damping to kill a small peak.** Sometimes a minor peak is better tolerated than smothered. Weigh what the fix costs the rest of the sound.
> - **Not matching left and right changes.** Every change you make to one channel, make identically to the other, and confirm it on the overlay.

What's Next

Damping is the acoustic half of tuning; the electronic half, correcting the residual shape in the signal itself, is covered in tuning with EQ (Chapter 23). And frequency response is the headline, but it isn't the whole picture. Decay plots, impulse response, and distortion measurements reveal problems that a frequency-response curve alone hides: the advanced measurement topics that round out this section and explain why two headphones with nearly identical curves can still sound clearly different from each other.

CHAPTER 23

Tuning with EQ

Where physical damping ends, EQ begins: how to correct the residual frequency-response errors in your build with parametric EQ, what EQ can and can't fix, and why you fix resonances acoustically first.

9 min read   makerphones.com/learn/tuning-with-eq

You've tuned with damping as far as it sensibly goes, and the curve is close but not quite: a broad dip here, a little too much energy there. EQ closes that gap. It's the most accessible tuning tool you have, and the easiest to misuse, because it fixes some problems completely and others not at all. Knowing which is which is the whole point of this chapter.

What EQ actually does

Equalization adjusts the level of specific frequency ranges in the signal before it ever reaches the driver. Parametric EQ, the kind worth using, lets you place a filter at a frequency you choose, set how much to cut or boost, and set how wide a band it affects (the Q). You're reshaping the frequency-magnitude curve electronically instead of acoustically.

Where damping changes the physics inside the cup, EQ changes the signal going into it. The curve can end up looking identical either way. But "looking identical" is exactly where people get into trouble, because the two are not the same underneath.

What EQ fixes well

EQ is genuinely good at broad tonal balance. If your build measures a touch bright, or the bass shelf is a couple dB light, or there's a gentle dip through the lower treble, a few wide parametric filters will bring the magnitude response right where you want it. It's also how you land precisely on a target after you've gotten close acoustically, which is exactly what tools like AutoEq automate, computing filters from a measurement to a chosen target. (Using those databases and targets is what the benchmarking `(Chapter 24)` chapter is about.)

For getting the overall tonal shape onto a target, EQ is fast, reversible, and free. Damping can't be undone once you've over-stuffed a cup. An EQ filter is a slider you can pull right back.

What EQ can't fix — and this is the important part

EQ moves the magnitude curve, but it does not change the physical behavior underneath it. Nobody tells you that part.

A **resonance** is the clearest example. You can pull the *level* of a resonant peak down with a narrow filter, and the magnitude curve will look flat right there. But the resonance is still a resonance: it still stores energy and rings after the signal stops, which you'd see on the decay plots in advanced measurement topics `(Chapter 25)`. A sharp resonance EQ'd flat in magnitude can still sound wrong. That's why the order is acoustic first, then electronic: you fix resonances with damping, not with a filter.

Distortion and breakup are the same story. If a driver is being pushed past its limits, EQ can't clean that up – turning a band down a little doesn't restore linearity, and turning it up makes things worse.

A **broken seal or a comfort problem** can be papered over on the graph, EQ a seal-related bass loss back up and the curve looks fine, but the seal is still leaking and still unstable from one wearing to the next. Fix the mechanical cause.

Channel imbalance from mismatched cups is the last trap. You *can* EQ each channel separately, but if your two builds don't match mechanically, you're chasing a moving target every time the seal shifts. Match them at the bench first, then EQ the pair together.

EQ corrects frequency-response magnitude. Ringing, distortion, leaks – none of those are magnitude problems, so EQ is the wrong tool.

Acoustic first, EQ last

So the workflow that actually works puts EQ at the end. Get the structure, the seal, the chamber, and the damping right, so the headphone is physically good on its own. *Then* reach for EQ to clean up the residual broadband tilt and settle precisely onto your target. Every time I've leaned on heavy correction to flatten a rough build, it's come out worse than a build that measured well and just needed a little polish.

Doing it without wrecking the signal

A couple of practical points. Boosting a band raises the overall level and can clip your source, so favor cutting over boosting where you can, and when you do boost, pull the overall preamp or gain down to leave yourself headroom. Apply your filters, then *remeasure with the EQ engaged* so you're checking what the build actually does, not what the filter math predicts. And as always, the graph serves the sound: if a filter measures right but sounds wrong, trust your ears, same as with damping.

One more thing worth keeping straight. EQ lives in your source: the player, the operating system, a little DSP box. It doesn't travel with the headphone. Plug into a different device and your correction is gone. That's perfectly fine for personal listening; just know the EQ isn't *in* the headphone the way damping is.

Common Mistakes

> ⚠ CAUTION
>
> - **EQ-ing down a resonance and calling it fixed.** The peak's level drops; the ringing stays. Fix sharp resonances with damping first, then EQ whatever's left.
> - **Big boosts that clip.** Boosting eats headroom and can drive your source into distortion. Cut instead of boost where you can, and use a negative preamp when you must boost.
> - **EQ-ing on a bad measurement.** A correction is only as trustworthy as the curve underneath it. Re-seat and remeasure before you build a filter on a feature that might be an artifact.
> - **Chasing a flat graph until it sounds dead.** The same trap as over-damping. Magnitude-flat isn't the goal, sounding right is.
> - **Forgetting it doesn't travel.** The headphone itself is unchanged. Anything you need baked in has to be done acoustically.

What's Next

EQ is the electronic half of tuning; tuning with damping `(Chapter 22)` is the acoustic half, and you'll almost always use them together: damping for the physical problems, EQ for the residual shape. To see the time-domain problems

EQ can't touch, advanced measurement topics (Chapter 25) covers decay and distortion plots. And both understanding frequency response (Chapter 2) and the benchmarking (Chapter 24) chapter cover the targets you're aiming all of this at.

CHAPTER 24

Benchmarking Your Build with Public Measurements

How to use public headphone measurement databases, oratory1990, Crinacle, squig.link, AutoEq, to benchmark your build and learn what good sounds like, and why your budget rig's numbers won't match theirs.

9 min read   makerphones.com/learn/benchmarking-with-public-measurements

You don't have to design in the dark. Some of the most useful headphone data in the world is free and public, thousands of commercial headphones measured carefully on reference rigs, and you can use all of it to figure out what you're aiming for and how close you've gotten. The catch is knowing how to compare honestly, because the numbers off your bench won't line up with theirs, and that's not a mistake on your end – it's just two different instruments measuring the same headphone.

What's actually out there

Over the last decade a handful of people quietly built public measurement libraries that have become the reference for the entire hobby.

- **oratory1990** measures over-ears and IEMs on a GRAS rig, the gold-standard hardware, with results referenced to the Harman target. It's widely treated as the most trustworthy single source.
- **Crinacle** runs one of the largest public databases anywhere, especially deep on IEMs, with interactive graphs you can overlay.
- **squig.link** is a whole family of those graphing tools, hosted by different reviewers, all letting you stack measurements on top of each other to compare.
- **AutoEq** is open source. It aggregates measurements from oratory1990, Crinacle, and others and computes the EQ needed to hit a target. Even if

you never touch the EQ, it's a clean way to see how a real measurement relates to a target curve.

What nearly all of them measure is frequency response, the same graph you've been learning to read. So you already have the skill. What you don't have yet is the habit of using theirs.

Why this is a gift for a builder

A public library lets you do a few things you can't pull off alone at a bench.

You can **learn what good looks like.** Scroll through measurements of headphones people love, then ones they don't, and the patterns start surfacing on their own: the gentle bass lift, the rise through the presence region, the way a harsh headphone shows a spike where your ears already told you it would be. You're training your eye against a few thousand data points instead of your own two or three builds.

You can **borrow a voicing you admire.** Find a headphone whose sound you love, or one you wish you could afford, pull up its shape relative to the target, and use *that* as your build's goal instead of guessing what "good" means. You're reverse-engineering a tuning that already won people over.

And you can **set honest expectations.** Look at what real bass extension looks like for an open-back versus a sealed design, and you'll stop chasing sub-bass your open build was never going to make. The data tells you what's physically on the table for the kind of headphone you're building.

The one trap: you can't compare the raw numbers

Here's what trips up nearly everyone the first time. oratory1990's curves come off a GRAS rig with a calibrated ear simulator: a precision coupler built to an international standard, designed to mimic the acoustics of a real ear and canal. Your budget rig doesn't have that. It's a different cavity, a different mic, and no standardized ear. So your raw curve and their raw curve of the *exact same headphone* will look meaningfully different, especially up high, past a few kHz, where the ear-canal resonance that a proper simulator models just isn't present the same way on your setup.

Your rig isn't wrong — it's just a different instrument, measuring a slightly different thing. (The measurement chapters go deep on why couplers matter; this is the practical consequence of all that.)

So do not lay your curve over an oratory1990 graph and try to match it decibel for decibel. You'll chase ghosts up in the treble for a week and learn nothing.

What you *can* compare is **shape and relative balance**: how far the bass sits above the mids, where the big peaks and dips land, the overall tilt from low to high. Those survive the difference in rigs reasonably well. Better still, compare **deltas to a target**: their headphone's distance from the Harman curve, and your build's distance from that same curve measured on your own rig. Two error curves are far more comparable than two raw curves, because each one already has its own rig's signature partly subtracted out.

The workflow I use

1. **Pick a reference.** Find a headphone in the database with a voicing you're after, and match the form factor to your build, because an open-back and a sealed can won't share a bass shelf. Note its shape relative to the target.
2. **Set your target.** Use that shape, or the Harman target itself, as your goal. The frequency-response chapter covers what the target means and where it came from.
3. **Measure your build on your own rig**: the same way, the same seating routine, every single time. Consistency on your bench matters more than agreement with theirs.
4. **Compare shapes and distance-from-target, never raw level.** Where's your bass relative to your mids? Where are your problem peaks? How does that compare to the reference's relationship to the same target?
5. **Tune toward the shape, re-measure, repeat.** Damping is how you move the curve once you know which direction it needs to go.

The database tells you where good headphones land; your own rig tells you where yours lands. Keep both honest and the gap between them is the work.

Common Mistakes

⚠ CAUTION

- **Matching absolute decibels across rigs.** Your 8 kHz peak and theirs aren't measuring the same thing. Compare shape and distance-from-target. Never raw level. This is the single biggest mistake, and it sends people down rabbit holes that don't exist.
- **Treating one measurement as gospel.** Even on a GRAS rig, a single seating shifts the result, and unit-to-unit variation between two copies of the same headphone is real. The good public measurers average several seatings; your numbers deserve the same respect before you trust them.
- **Chasing an exact copy of a database curve.** It was measured on different hardware, on a different physical unit, maybe with different pads than yours. Aim for the overall shape, not an exact match to their line.
- **Forgetting it's data, not preference.** A target tells you what most listeners prefer on average. It doesn't tell you what *you* prefer. Use the databases as a starting line, then trust your own ears to make the last calls.

What's Next

If those graphs still look like noise, go back to understanding frequency response (Chapter 2). Everything here is built on reading that shape, and on the Harman target it explains. To get your own measurements trustworthy enough to compare against anything, taking and interpreting measurements (Chapter 20) and the budget measurement setup (Chapter 19) are the foundation, and tuning with damping (Chapter 22) is how you actually move your curve once the data shows you where it needs to go. The sources and further reading page links straight to the databases themselves.

CHAPTER 25

Advanced Measurement Topics

Beyond frequency response: decay plots, impulse response, and distortion measurements, what they reveal, and which ones a DIY builder actually acts on.

10 min read   makerphones.com/learn/advanced-measurement-topics

Frequency response is the headline measurement: it tells you most of what you need while you're tuning. But it's a snapshot. It tells you how loud each frequency is and nothing about how long each one rings, and it says nothing about distortion at all. That's why two headphones with nearly identical curves can still sound clearly different from each other. The measurements in this chapter fill in those gaps. You won't need all of them, but a couple are genuinely useful, and it's worth knowing what the rest are for so you can read them when you see them.

The time dimension: decay plots

A frequency response curve has no time axis. It can't tell you that a particular frequency keeps ringing after the signal stops, and that ringing is a big part of what you hear as hardness, smear, or one-note bass.

A decay plot, you'll see it called cumulative spectral decay, a waterfall, or a CSD, adds that missing axis. Picture the familiar frequency-response curve as the front edge, then a series of fainter copies stacked behind it, each one showing the response a fraction of a millisecond later as the energy dies away. Frequencies that stop cleanly drop straight down into the floor; frequencies that ring keep standing tall in those later slices, forming a ridge that runs back into the plot. Those ridges are resonances, and their height and length tell you how badly each one rings.

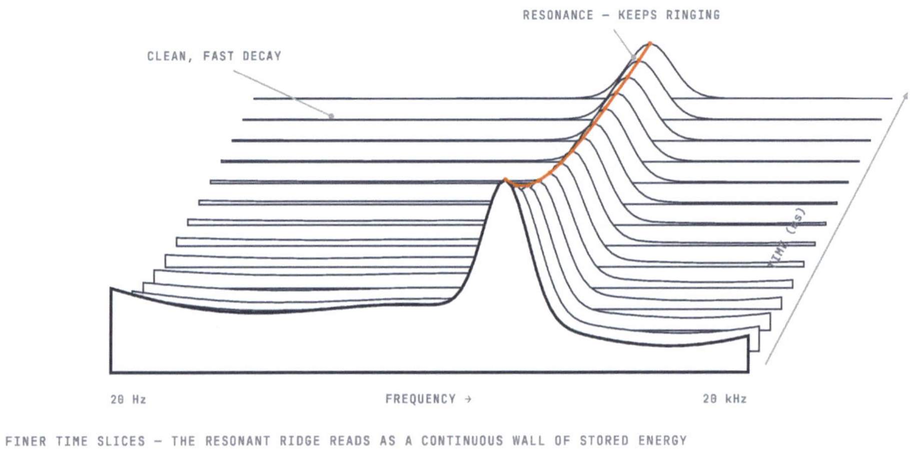

Fig. 25.1 – A decay waterfall: the ridge that keeps standing is a resonance ringing.

This is the single most useful "advanced" plot for a builder, because it shows you exactly which frequencies are misbehaving, the ones to chase with damping or a structural fix, and it shows you whether your fix worked. A clean, fast decay is the tight, controlled sound you're after. Long tails are the smear and fatigue you're trying to get rid of.

Impulse and step response

The impulse and step responses show the driver's raw reaction to a sharp input in the time domain: how it starts and how it stops. A clean impulse rises sharply and settles quickly back to the line; a messy one wobbles, overshoots, or shows a delayed second bump. That second bump is the tell you care about: it usually means a reflection or resonance bouncing around inside the cup and arriving late, the same problem the decay plot shows from a different angle.

For day-to-day tuning you'll act on these rarely, the decay plot presents the same information in a form that's easier to act on, but they're the raw material the decay plot is built from, and they're good for spotting gross timing problems at a glance.

Distortion

A real driver isn't perfectly linear: it adds harmonics that weren't in the original signal, and a distortion plot shows how much, across frequency, usually as a percentage. For a decent driver at normal listening levels, distortion is low and inaudible. It tends to climb in the deep bass, where the diaphragm moves the most and works hardest, and at high volume, where the driver is being pushed near its limits. Seeing it rise in those two places is normal and expected.

For a builder, though, a distortion plot earns its keep less as a way to grade the driver and more as a way to catch faults. Different faults leave different signatures. A buzz from a loose part or a slack wire tends to throw up sharp distortion spikes at specific frequencies. A rubbing or bottoming driver smears distortion across a broad range. An air leak can show up as noise or a whistle riding on top. The common thread: when distortion jumps somewhere it has no business jumping, or turns rough where it should be smooth, stop tuning and go look for a physical cause. A distortion plot is the best lie-detector you have for "is this a tuning problem or did something come loose."

Spectrograms and the rest

A spectrogram gives you a fuller time-and-frequency picture, frequency on one axis, time on another, intensity as color or brightness, and it's a nice way to see resonances and reflections laid out at once. It's informative and entirely optional. Treat it, and the other exotic plots your measurement software can produce, as things worth understanding so you're not lost when you see them. They're not part of your regular workflow.

What actually matters for DIY

For the vast majority of builds, frequency response plus a decay plot covers nearly everything you can act on, and an occasional distortion sweep catches gross faults. The rest are worth knowing about and rarely worth running. Don't let collecting impressive-looking graphs replace acting on the two that matter. If you're not changing anything based on a plot, you're just collecting graphs.

Read them comparatively

The same honesty from earlier applies, doubly. A hobby rig measures the driver and the rig together, so absolute decay and distortion numbers aren't trustworthy, and the high treble is especially noisy across all of these plots. Use them comparatively, before versus after a change, left versus right, this build versus that one, and never as an absolute score. And remember that decay and distortion both shift with seating, so re-seat and re-run before you believe a small feature is real.

> TIP
>
> When a decay plot shows a stubborn ridge at one frequency, that's your target: note the frequency, apply a fix (damping if it's a cavity, stiffening if it's structural), and re-run the plot to watch the tail shorten. Seeing the ridge collapse is the most satisfying confirmation in the whole hobby that you fixed the right thing, rather than just moved a bump on a curve.

Common Mistakes

> CAUTION
>
> - **Reading absolute distortion numbers as gospel.** A hobby rig's absolute figures aren't reliable. Compare, don't grade.
> - **Ignoring a distortion spike that's really a rattle.** A sharp spike often means a mechanical fault, a loose part or a leak, not a voicing issue. Go find the physical cause.
> - **Chasing top-octave decay wiggles.** Tiny tails up high are usually rig artifacts. Re-seat and re-run before acting.
> - **Collecting plots you don't act on.** Impressive graphs aren't progress. Frequency response plus a decay plot is where the action is.
> - **Forgetting that seating moves everything.** Decay and distortion both change with position. Establish a repeatable baseline first.

What's Next

These plots are how you *find* resonances: the ringing ridges on a decay plot are resonances made visible. The next chapter is about controlling them at the source, in the driver, the cavities, and the structure of the headphone itself: resonance control (Chapter 27).

PART

5

Advanced

CHAPTER 26

Acoustic Modeling

Simulating how a driver and chamber will behave before you build: the lumped-element approach, the tools a hobbyist can actually use, and why a model never replaces measuring.

11 min read   makerphones.com/learn/acoustic-modeling

Every chapter so far has worked the same way: design by principle, then confirm by building and measuring. Acoustic modeling adds a step in front of all that. Instead of printing a cup to discover how it sounds, you simulate how a driver and chamber will behave first, narrowing the design space and, just as valuably, building your intuition for why a change does what it does. Modeling won't replace building and measuring, and a model you trust too much will lead you astray, but used well it can save you real iterations and deepen your understanding of everything in the earlier chapters.

What acoustic modeling actually is

The core idea is the equivalent circuit. A headphone's driver and the air around it form a single electro-mechanical-acoustical system, and that system can be modeled as an analogous electrical circuit. The driver's electrical, mechanical, and acoustic behavior becomes circuit elements, and so do the chambers: the springiness of trapped air becomes a compliance, the air in a vent or port behaves like a mass, and damping and leakage act as resistances. Solve the circuit, and you get a predicted frequency response. It is, quite literally, SPICE for acoustics.

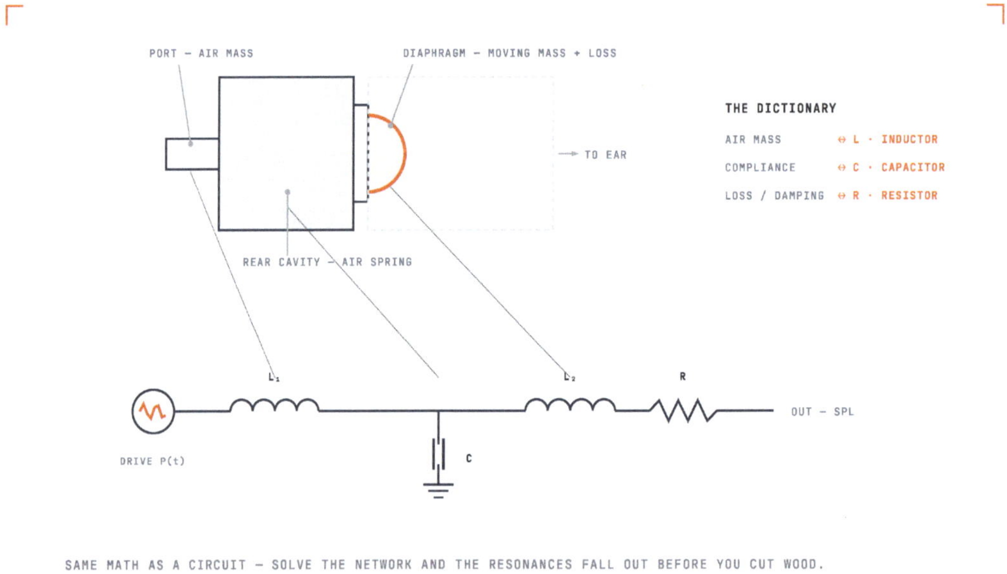

Fig. 26.1 – The lumped-element analogy: driver, chambers, and port drawn physically, then as the circuit they behave like: air mass as inductance, compliance as capacitance, loss as resistance.

This is the lumped-element method, and it's the tractable, fast approach that suits a headphone well, because the chambers are small relative to the wavelengths involved. It handles the things you care about, how the rear volume, a vent, and damping interact to shape the bass and the main resonance, without needing a supercomputer.

There's a heavier tier above it. Boundary-element and finite-element methods solve the wave equation across a 3D mesh of your geometry, capturing radiation, diffraction, and spatial detail that the lumped model can't. They're far more accurate for those effects and far more demanding to set up, and the genuinely capable tools are either expensive or hard to learn.

The parameters you need

A model is only as good as the driver data you feed it, and the data that matters is the set of Thiele-Small parameters, the driver's key electro-mechanical numbers: its resonance, its compliance, its moving mass, the strength of its motor, and so on. These are what the equivalent circuit runs on.

DIY drivers often ship with incomplete or optimistic datasheets, so the reliable move is to measure your own parameters rather than trust the sheet. You can do this with the same measurement gear from the last section, an impedance measurement in REW, combined with the added-mass or known-volume method, to extract real numbers for your specific driver.

The tools you can actually use

For the accessible, lumped-element tier, two tools dominate. **Hornresp** is free, widely used, and despite its name handles sealed and vented systems well enough to approximate headphone chamber behavior. **AKABAK** is a professional electroacoustic simulator that couples lumped-element and boundary-element methods; it offers a free version that's essentially the full program but can't save result files, which is plenty for learning and exploring. There are also open-source equivalent-circuit tools floating around for the curious.

A step up in the speaker world, **VituixCAD** is free and excellent for crossovers and diffraction, more loudspeaker-oriented than headphone-specific but useful background. At the top sits full finite-element multiphysics software like **COMSOL**, which can model the voice coil and cavities in exquisite detail, and which is too expensive for hobbyists. The honest state of things is that 0D/1D lumped modeling is in good, accessible shape, while the 2D/3D world remains the professional frontier.

Using it for a headphone

The practical workflow is to model the driver and the front and rear cavities as a lumped network, then sweep the variables you'd otherwise have to print to test, the rear volume, the venting, the damping resistance, and watch the bass and the resonances move on screen. This turns "print it and find out" into "predict it, then print to confirm." You learn, in an afternoon and for free, how sealed and vented behavior differ for your driver and how volume shifts the resonance, before you commit a single gram of filament.

Use the model to pick a sane starting chamber and to see which levers matter. It won't hand you the final tuning; the bench finishes that job.

The honest limits

Headphone acoustics are genuinely hard to model accurately, and it's important to know that going in. The cavities are tiny and coupled, the load your ear or coupler presents is complex and varies from person to person, and leakage, the pinna, and the seal all matter and all resist tidy modeling. A model captures trends and gets you into the ballpark; it will not match reality precisely.

A model gets you close; the bench gets you right. Simulate to narrow things down and understand them, then prototype and measure to confirm and tune. When a careful measurement disagrees with a model, the measurement wins. I don't know anyone good who skips this step. The screen lies sometimes; the coupler doesn't.

> TIP
>
> Before you trust a model, measure your driver's actual Thiele-Small parameters rather than relying on a DIY driver's datasheet. Those sheets are often incomplete or optimistic, and a model fed bad parameters will confidently predict the wrong thing. REW with a known resistor and the added-mass method gets you real numbers to simulate from, and a model built on real data is the difference between insight and fiction.

Common Mistakes

> CAUTION
>
> - **Expecting the model to nail the final tuning.** It narrows and explains; it doesn't finish the job. The bench does that.
> - **Feeding it unverified datasheet parameters.** Measure your driver's real Thiele-Small numbers, or the prediction is built on sand.
> - **Ignoring the ear load and leakage.** A simple model handles these crudely, and they're a big part of the real result. Treat the output as approximate.
> - **Reaching for FEM when lumped modeling answers the question.** The heavy tools are overkill for most chamber questions, and COMSOL is out of reach anyway.
> - **Skipping the build-and-measure confirmation.** A model is a guess until you build it and measure. Confirm before you trust it.

What's Next

With a way to predict behavior before you build, the natural next step is controlling the resonances that both your model and your measurements will reveal, in the driver, the cavities, and the structure. That's resonance control (Chapter 27).

CHAPTER 27

Resonance Control

Where resonances come from in a headphone, driver, cavities, and structure, and the four levers you have to control them. The deeper skill underneath tuning.

11 min read   makerphones.com/learn/resonance-control

Every chapter so far has circled the same enemy. The peak you damped, the boxiness you chased, the ring you saw on a decay plot, the one-note bass, those are all resonance, showing up in different places under different names. This chapter names it directly, because once you understand resonance as a single phenomenon with a handful of sources and a handful of cures, a lot of headphone behavior that seemed mysterious turns into something you can reason about.

What a resonance is

Any structure, and any volume of air, has natural frequencies where it wants to vibrate or oscillate. Feed energy in at one of those frequencies and it piles up rather than passing through, and it keeps going after the signal stops, that's the ringing tail you saw on a decay plot. A resonance is just that: a frequency where energy accumulates and rings instead of being delivered cleanly and then stopping.

In a headphone you don't have one resonance, you have several, stacked on top of each other from different sources. Untangling which is which is most of the work.

Where they come from

The driver itself. Every driver has a fundamental resonance down low, its natural "want to move" frequency, and, higher up, breakup modes, where the diaphragm stops moving as a single piston and different parts of it flap out of step with each other. You're not going to redesign the driver, but how well-behaved these are is part of why one driver sounds cleaner than another, and it's a reason to choose carefully: see driver technologies (Chapter 6). You design around a driver's resonances rather than erasing them.

The air cavities. The front and rear chambers have their own resonances and standing waves, set by their size and shape. These are the ones acoustic chamber design (Chapter 15) and damping address most directly: the boom, the boxiness, the cupped coloration.

The structure. The cup walls, the baffle, the headband. This is the source DIY builders most often overlook, because it isn't part of the "acoustics" mental model, but a thin printed cup wall rings like a drum head, a flimsy baffle flexes and adds its own voice, and a resonant headband transmits buzz straight into the cup. A surprising amount of "this build sounds hard and cheap" is actually the plastic shell ringing along with the music.

Why it matters

Each resonance shows up as a peak and a ringing tail. That means coloration, a honk, a one-note region, a hard upper midrange, plus smeared detail and listening fatigue. Flattening the frequency response alone can't fully fix this, because a resonance is a time problem as much as a level problem. A flat headphone can still sound smeared and tiring if its resonances ring, and fixing the ringing is what clears that up.

The four levers

Absorb it (damping). Convert the resonant energy to heat with felt, foam, or fill. This is your main tool for cavity resonances and reflections, and it's covered in depth in damping strategy and application (Chapter 17). Absorption is the right answer when the resonance lives in the air.

Stiffen it or add mass. This is the right answer when the resonance lives in the structure. Making a panel stiffer raises its resonant frequency, often pushing it up out of the range where it matters, and reduces how much it rings. Adding mass lowers the frequency and deadens it. For a printed cup, the most direct fixes are increasing wall thickness, adding internal ribs, or bumping up the infill, all of which turn a thin ringing shell into a dead one.

Break up the geometry. Parallel walls and simple box shapes encourage standing waves, because the sound bounces back and forth between two flat surfaces and reinforces itself at specific frequencies. Non-parallel surfaces, chamfers, and irregular internal shapes spread that energy out instead of letting it build at one frequency. You can design a resonance down before you ever damp it.

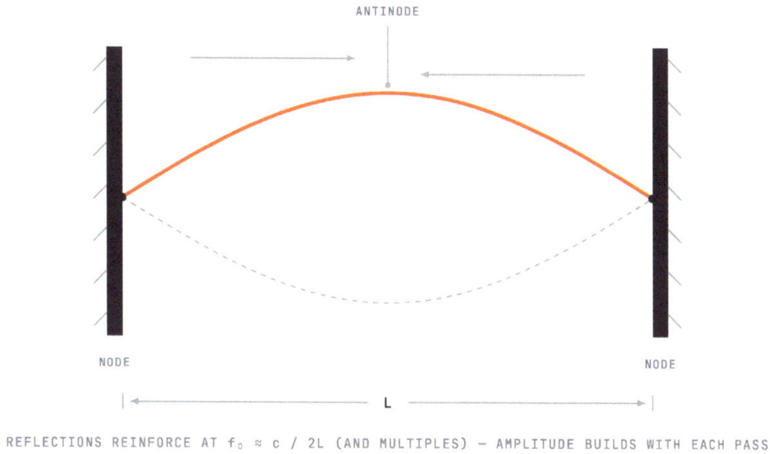

Fig. 27.1 — A standing wave between parallel walls: energy building at the cavity's resonant frequency.

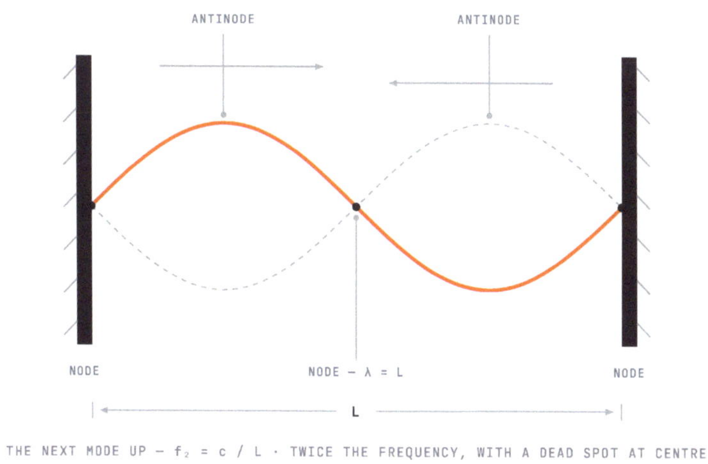

Fig. 27.2 — The next mode up: the second harmonic rings at twice the frequency, with a dead spot at center.

Decouple it. Stop vibration from traveling between parts. Isolating the cup from the gimbal and headband keeps structural buzz from reaching the cup; soft-mounting the driver keeps its frame from driving the shell.

There's a fifth, worth a mention: constrained-layer damping, where a stiff layer and a soft lossy layer are bonded to a panel so that any flexing shears the lossy layer and burns off the energy. It's an efficient way to kill panel ringing in a stubbornly resonant cup wall.

Putting it together for a printed headphone

The three sources map cleanly onto three responsibilities. The chamber design sets your cavity resonances, so design them well. Damping controls the cavities and reflections, so apply it deliberately. And the structure controls the shell, so manage it with wall thickness, ribbing, material choice, and decoupling.

The classic DIY failure is to attack a structural resonance with absorption: to keep adding felt to a cup that's actually ringing in its walls. No amount of felt fixes a structural resonance, because the energy is in the plastic, not the air. When damping isn't working, ask whether you're damping the wrong kind of resonance, and reach for stiffness or decoupling instead.

Finding them

A decay plot, from advanced measurement topics (Chapter 25), is your map: it shows which frequencies ring and for how long, which tells you what to control and whether your fix actually shortened the tail. Resonance control without a way to see decay is possible, but it's a lot of guessing.

> TIP
>
> Tap the empty cup gently and listen. A sharp, ringing "ting" means a resonant shell you'll hear coloring the music; a dull, dead "tock" means a well-behaved structure. It's the crudest possible resonance diagnostic and it costs nothing, and it instantly tells you whether your next move should be more damping or a stiffer wall.

Common Mistakes

⚠ CAUTION

- **Damping a structural resonance.** If the cup walls are ringing, felt won't fix it. Stiffen, mass-load, or decouple instead.
- **Leaving thin printed walls to ring.** A thin shell is a drum. Add thickness or ribs if it sounds hard and the tap test rings.
- **Ignoring driver breakup when choosing a driver.** How clean a driver stays at the top is part of what you're buying. Choose with it in mind.
- **Parallel-walled box cavities.** Flat parallel surfaces build standing waves. Break up the geometry where you can.
- **Over-stiffening everything.** Mass and ribs add weight and don't help if the real problem was a cavity. Match the lever to the source.

What's Next

Controlling resonance reliably in one build is a skill; controlling it the same way across many builds is a process. Manufacturing for consistency (Chapter 28) is about making your results repeatable, matched, reproducible, and shareable, rather than lucky.

CHAPTER 28

Manufacturing for Consistency

Moving from a lucky one-off to a repeatable design: matched channels, reproducible builds, and the documentation that lets you (or anyone) build it again.

11 min read makerphones.com/learn/manufacturing-for-consistency

Plenty of promising DIY builds sound wonderful exactly once, on one head, and never again. The jump from that to a design is what this chapter is about: matched left and right, reproducible from one batch to the next, and pinned down well enough that someone else could build it and hear what you heard. Consistency is the line between a project and a design.

Why consistency matters

Three things break without it.

First, channel matching. Left and right have to be close to identical or the stereo image suffers: it drifts, smears, or sits off-center. You saw this when you overlaid the two channels while tuning. Consistency is what keeps a matched pair matched.

Second, reproducibility. You tuned a build over many evenings of small changes. Can you build a second unit that sounds the same? Can you build it again in six months, after you've forgotten the details? If the answer is no, what you have is a one-off.

Third, shareability. The moment you want to publish a design so others can build it, your result only transfers if the design specifies everything that matters. A vague recipe gives everyone a different headphone.

Where variation creeps in

Printing. Printer calibration drifts over time, filament varies by brand, batch, and even color, temperature and humidity shift how parts come out, nozzles wear, and orientation changes both dimensions and how a wall behaves. Two prints from the "same" file weeks apart are not guaranteed to be the same part.

Drivers. Units vary straight out of the box. Two nominally identical drivers can differ enough to hear, and the cheaper the driver, the wider that spread tends to be.

Assembly. How much damping went in: a pinch versus a measured amount. How good the seal is. How far the gasket compressed. Where exactly the felt ended up. Each is a small variable, and they stack.

Time. Foams compress, pads wear, adhesives age. A build drifts even after it's finished.

Controlling print consistency

Lock down a calibrated, documented print profile and reuse it exactly: same settings every time, written down where you can't fudge them from memory. Use the same material and, for a matched pair, ideally the same filament batch. Keep orientation consistent. And verify the parts that matter with calipers, because the model is not the part: your printer has real tolerances and shrinkage, and a dimension that's critical to a seal or a fit needs to be measured on the actual print, not trusted from the CAD.

Matching drivers

For a pair where matching matters, which is most over-ear builds, measure and pair your drivers, or at minimum source them from the same batch. Again, the cheaper the driver, the more this matters, because the spread is wider. A few minutes matching a pair saves you chasing a channel imbalance that no amount of damping will fix, because it was never an assembly problem.

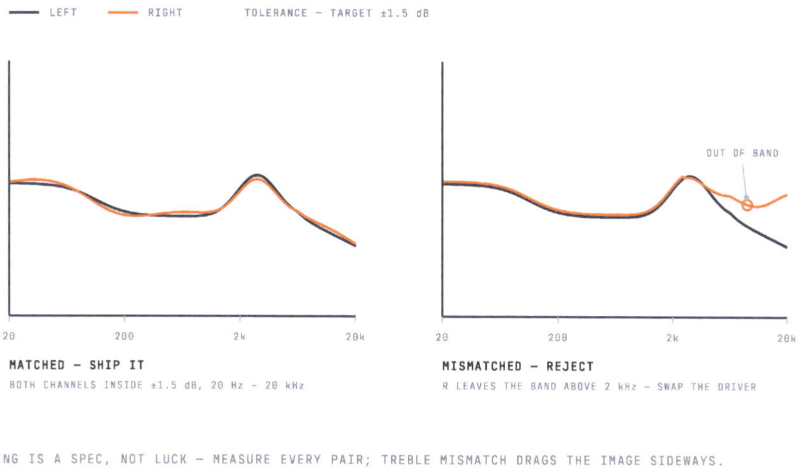

Fig. 28.1 – Channel matching against a ±1.5 dB tolerance band: a matched pair ships; a right channel that leaves the band above 2 kHz means swapping the driver, not reaching for damping.

Controlling assembly

Write the process down and follow it the same way each time. Weigh your damping fill so both cups get exactly the same amount: this is the single easiest consistency win available to you, and it turns "stuff some in there" into a number you can repeat. Compress gaskets the same. Build both channels in the same session, with the same materials, in the same way. Once you're building more than a couple of units, simple jigs or fixtures for repeatable placement start to pay off.

Documentation is the actual deliverable

Here's the part most builders skip. Print files and a parts list aren't enough on their own. What makes a design reproducible is a complete build spec. Dimensions and print profile. The damping recipe, by weight, by location. The exact driver. The gasket. The assembly steps and the order. "Some felt in the back" is not a design; "this much fill behind the driver, this lining on the walls" is. This is also what turns a build you'd like to share into something others can actually replicate rather than approximate. Treat the document as the real output and the physical headphone as proof it works.

Quality control

Measure every finished unit. Overlay left against right, and compare both against your reference for the design. A quick end-of-build measurement catches the unit that drifted, the gasket that didn't seat, the fill that was off, the driver that wasn't as matched as you thought, before it leaves your bench. Rework or set aside the outliers. Five minutes with the rig at the end is cheap insurance against shipping or keeping a build that quietly went wrong.

The mindset

This is really a single shift in what "done" means. Done isn't "it sounds good." Done is "I can make this again and get the same headphone." Hold yourself to that, and the lucky accidents stop.

> TIP
>
> Weigh your damping. A small kitchen scale that reads to a tenth of a gram turns "stuff some fill in there" into a repeatable recipe, and matched fill weights between left and right is the cheapest, most reliable channel-matching you will ever do. It's a few dollars of tool that fixes a problem people spend whole evenings chasing.

Common Mistakes

> ⚠ CAUTION
>
> - **Never writing down what you did.** A one-off you can't reproduce isn't a design. Document the build spec as you go.
> - **Eyeballing damping amounts.** Weigh them. Guessed fill is the most common source of channel mismatch and batch drift.
> - **Mismatched filament or driver batches in a pair.** Source matched, especially for cheaper drivers, where the spread is wide.
> - **Trusting CAD dimensions over the printed part.** Measure critical dimensions with calipers. The model is not the part.
> - **Skipping the end-of-build QC measurement.** A quick overlay catches the unit that drifted before it gets out the door.

What's Next

With consistency handled, the last piece of the advanced material is the judgment that turns good specs into a headphone people actually love, the things the industry teaches that no measurement quite captures. That's professional design insights (Chapter 29).

CHAPTER 29

Professional Design Insights

The judgment that separates a good prototype from a headphone people keep, lessons from years in the industry that no spec sheet or forum thread quite captures.

11 min read   makerphones.com/learn/professional-design-insights

Every chapter so far has been technique: drivers, chambers, damping, resonance, measurement. This one is about judgment: the things that twenty-five-plus years around headphones taught me that no spec sheet or forum thread quite says out loud. I came up through this industry, early days at Grace Design, then a long run at HeadRoom, and these days designing my own open-back headphone, the Daily Driver, in the open, and most of what follows is the stuff I had to learn the expensive way. I'm offering it so you can skip some of the tuition.

None of this replaces the technique in the earlier chapters. It's the layer on top. Technique gets you a working headphone. Judgment is what makes it one you keep reaching for.

A prototype is the easy twenty percent

Making something sound great once, on your own head, is the fun part, and it's the small part. The hard eighty percent is making it sound the same on the next unit, and the next hundred: the tolerances, the pad that wears, the seal that varies from head to head, the driver that came from a different batch. The industry learns this the hard way, usually right after it falls in love with a prototype. You can learn it cheaply by respecting it early, which is the whole reason the consistency chapter exists.

The seal and the pad *are* the sound

More than you would believe. A huge fraction of what people call a headphone's signature is actually the pad and the seal, and both vary across heads, wear over time, and move the bass dramatically. I have watched a "voicing problem" turn out to be nothing but a tired pad. Treat the pad and the seal as first-class design elements, not afterthoughts, and don't chase a one-decibel tweak in the tuning when a pad swap moves things by five. Ear pads and comfort (Chapter 10) does as much acoustic work as anything in the design — the comfort is almost a side effect.

There is no free lunch — design is choosing trade-offs

Isolation versus openness and stage. Bass extension versus midrange purity. Comfort versus seal. Smooth versus detailed and vivid. Beginners try to maximize all of them at once and end up with mush, because these things genuinely pull against each other. Good design picks a coherent intent and trades deliberately toward it. Decide what this headphone is *for*, and who it's for, and how it'll be used, then trade in that direction on purpose. A headphone trying to be everything ends up vague, and people feel that even when they can't tell you why. Open vs closed back design (Chapter 4) is the first and biggest of these forks, but it's far from the only one.

Voice to an intent, not to a graph

A target curve is where you start, not where you stop. The best-sounding things I've been part of were voiced by trained ears toward a clear intent, with measurement keeping them honest, not by flattening a curve until the graph looked tidy. The graph and the ears each catch what the other misses: the graph sees the imbalance your ears adapt to, and your ears hear the life that a "correct" curve can lack. Trust neither one alone. The measurement chapters give you the discipline; your intent gives it a direction. See why measure headphones (Chapter 18) for where that line sits.

Simpler usually wins, and serviceable always does

Over-engineering is a trap. More parts means more variation, more cost, more ways to fail, and rarely better sound. The designs that last are often the simplest ones that do the job well, and the ones you can open up, fix, and iterate on. Design so that future-you can take it apart, because future-you will need to: to retune, to swap a driver, to chase down a rattle. Glue a clever assembly shut and you've sealed off your own ability to fix or retune it later.

The driver and the pad dominate both cost and sound

Everything else supports them. Spend your attention, and your budget, where it actually moves the needle. A beautiful cup wrapped around a mediocre driver is a beautiful disappointment, and no amount of careful damping rescues a driver that was the wrong choice. When you're deciding where to put effort, start from the two parts that determine most of the result and work outward.

What manufacturers actually optimize for

Visiting manufacturers and talking with their engineers taught me something: when you buy a commercial headphone, the acoustics are not the only thing that got optimized, and for most mass-market products they're not even the primary thing. The team balanced acoustics against manufacturing cost, assembly time, component yield, warranty return rate, packaging appeal, and competitive positioning against everything else at that price point. The acoustic engineer usually knows the headphone could sound better with different materials, and the margin doesn't support those materials. So it ships to hit the margin, even though the engineer knows it could sound better.

That's not cynicism, it's just how product development works at scale, and it's completely reasonable given the constraints. But it changes how you read commercial products: some of their design decisions are compromises you don't have to make. You're not building to a margin, you have no assembly-line clock, and you can use materials and methods that would be impractical in production. Copy the good decisions, not the compromised ones.

Every serious manufacturer measures — "by ear alone" is mostly mythology

The romantic idea that real audiophile designers work by ear and leave the test gear to the bean counters is, in my experience, backwards. Every serious manufacturer measures extensively, GRAS ear simulators, anechoic chambers, distortion rigs, and the listening sessions come *after* the measurements, not instead of them. The best work integrates both, which is exactly the discipline the measurement chapters (Chapter 18) are trying to hand you. Don't let anyone talk you out of the rig.

One thing that stays genuinely hidden, though, is damping. The specific combination of damping materials, their precise placement, and the reasoning behind those choices is some of the most proprietary knowledge in the industry. The ZMF damping patent this manual references is a rare public glimpse: most of it stays internal. The good news is that systematic measurement and iteration can reverse-engineer the *effects* even when the exact recipe stays locked in someone's lab.

Expect to build more prototypes than you think

If there's one number worth tattooing somewhere, it's this: serious headphone designs go through roughly `15-40 prototypes` before a version ships. The first one almost never sounds right, the fourth is closer, and the design the company ships might be revision twenty. So when a DIY builder gets frustrated and walks away after `3-4 iterations`, they're not failing, they're stopping early by professional standards. Version one is a draft whose whole job is to tell you what to fix in version two.

Where DIY builders actually win

For all that, you hold advantages no product company can touch. Your iteration is fast because cost isn't a variable in the loop: you can make ten damping changes over two weekends without a single margin meeting. You're fitting one specific head, so clamp, pad depth, and cup size can all be tuned for the person actually wearing them, which no mass-market product can do. You're free to chase walnut cups, odd driver pairings, and experimental damping, because the worst case is that you learn something and reprint. And you can document everything honestly: the tuning decisions and dead ends a company has to keep as competitive secrets, you can publish, and get feedback that makes the next build better.

Listen to people who live with it

Listen past the measurement, and past the loudest voice on the forum. What spec-sheets well and what people actually reach for every day are not always the same headphone. Real, long-term listening, is it comfortable for hours, is it fatiguing, do I keep coming back to it, tells you things a five-minute graph never will. Get your builds onto other heads and ask what they notice after a week, not after a minute. Forums chase whatever's fashionable. The person who just quietly keeps wearing your build — that's your real signal.

Stay humble — even after all those years, I get it wrong

I still misjudge things and iterate. Everyone does, at every level. Build many, measure, listen, kill your darlings, and stay curious. That isn't a beginner's disclaimer to be outgrown: it's the actual job, and the people who are best at it are the ones who never stopped treating it that way. You're learning in public, and so is everyone worth learning from.

> TIP
>
> Before you obsess over the graph, do the boring checks: is the seal good, are the pads fresh and matched, does the headphone fit *your* head right. The large majority of "mystery" sound problems are one of those three, and chasing them in the tuning, when the real cause is a worn pad or a poor fit, is how you burn a whole weekend solving the wrong problem.

Common Mistakes

> ⚠ CAUTION
>
> - **Maximizing every parameter.** Trying to win every trade-off at once gives you mush. Choose an intent and trade toward it.
> - **Chasing tiny measured tweaks while ignoring pad and seal.** Those variables dwarf the decibel you're hunting. Fix the big levers first.
> - **Over-engineering.** Simpler is usually better and serviceable is always better. Don't build what you can't open.
> - **Trusting the graph or the forum over long-term listening.** Live with it, and get others to. What people keep wearing is the real verdict.
> - **Calling it done because it sounded good once.** Good once, on your head, is a prototype, not a design.

What's Next

That's the close of the advanced material. From here the manual heads into special topics, adding wireless, noise cancelling, microphones, and the particular world of custom in-ear monitors, and into the hands-on build guides where all of this technique and judgment finally meets a workbench. Everything after this is practice: you keep putting builds on heads and paying attention to what comes back.

PART

6

Special Topics

CHAPTER 30

Bluetooth Integration

Adding wireless to your build: how Bluetooth audio works, the codec landscape, the receiver boards a DIY builder can integrate, and the battery-and-charging subsystem it all rides on.

12 min read   makerphones.com/learn/bluetooth-integration

Wireless is near the top of every "can I add this?" list, and the good news is that, unlike active noise cancelling, it's genuinely achievable in a DIY build. The catch is that Bluetooth is a whole subsystem — a receiver board, a battery, a charging circuit, and an antenna — not something you bolt onto a driver. This chapter maps how Bluetooth audio actually works, the codec landscape everyone asks about, and the realistic path to integrating it without designing wireless electronics from scratch.

How Bluetooth audio works

The chain is worth picturing end to end. Your source, a phone, usually, compresses the audio with a codec and transmits it. A Bluetooth receiver chip inside your headphone catches that signal and decodes it back into a standard digital audio stream. That stream feeds a DAC, which converts it to analog, which feeds an amplifier, which drives your driver. A Bluetooth headphone, in other words, is a small computer plus a DAC plus an amp plus a battery, all wrapped around the driver you built.

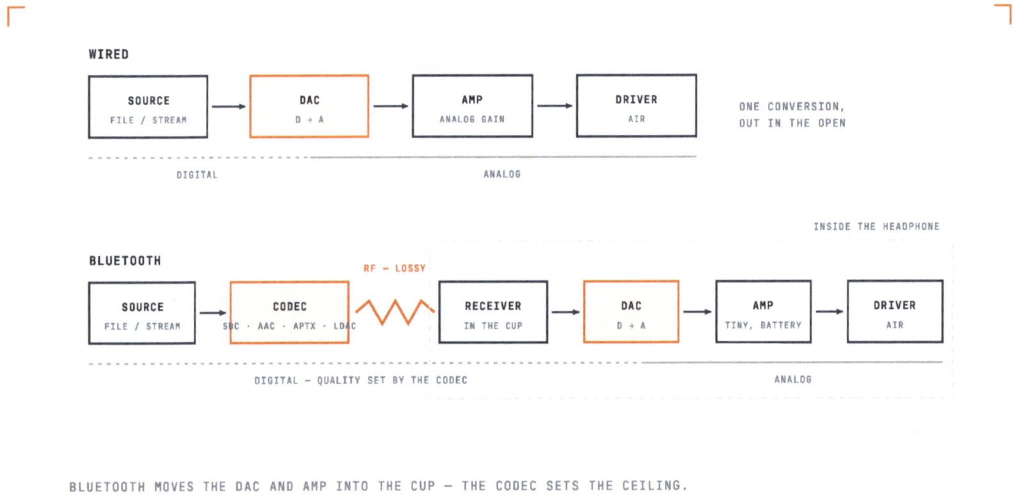

Fig. 30.1 – Wired vs Bluetooth signal chain: Bluetooth moves the DAC and amp inside the cup, and the codec sets the quality ceiling.

The codecs everyone asks about

Codecs are where most of the questions live, so here's the landscape as it stands.

SBC is the universal baseline. Every Bluetooth audio device supports it, it's the lowest quality of the bunch, and it's your guaranteed fallback when nothing better is shared. **AAC** is the standard on Apple devices and sounds good there. The **aptX family** from Qualcomm runs from plain aptX, through aptX HD for higher quality, to aptX Adaptive, which adjusts its bitrate to the connection and latency, up to aptX Lossless, which reaches CD quality. **LDAC**, Sony's codec, targets high-resolution audio at high bitrates, and **LHDC** is a comparable codec with narrower support. **LC3** is the newer codec built for Bluetooth LE Audio, more efficient than the old ones and the foundation for broadcast features like Auracast. LC3 is to LE Audio what SBC is to classic Bluetooth, and it's the direction the standard is heading.

Here is the rule that matters more than any quality ranking: a codec only works if *both* ends support it. A board boasting LDAC or aptX does nothing if your phone doesn't speak that codec. The connection falls back to SBC. So the codecs on the box are a ceiling, not a promise, and SBC is the floor that always works.

The DIY path: receiver boards

The realistic approach is not to design Bluetooth electronics but to integrate a pre-made Bluetooth audio receiver board, which takes the entire wireless-and-decoding job off your hands.

These come in two output styles. A board with **analog output** has a DAC, and sometimes a headphone amp, built in, the simplest option, since you can run it almost straight to your driver. A board with **I²S digital output** hands you the raw digital stream to feed into your own separate DAC and amp, more work, and more potential for quality if you care to chase it.

For chips, a few define the current landscape. The **Qualcomm QCC3034** is a common entry point. The **QCC5125** is the popular enthusiast choice, supporting the aptX family and, through community firmware, LDAC; inexpensive receiver boards built on it output I²S on solder pads and ship with an antenna. The newer **QCC5181** flagship adds LDAC, LHDC, and LC3. And fully integrated modules like the IOT747 IDC series deliver LE Audio with LC3 and Auracast alongside classic aptX and aptX Lossless on Bluetooth 5.3, in a tiny pre-certified package. Pick a board by the codecs you actually want and the output style you prefer.

The rest of the subsystem

The part builders forget is everything *around* the board. Bluetooth needs power, which means a battery, typically a small lithium-polymer cell, plus a charging and protection circuit. For charging, the canonical DIY part is the `TP4056` board: a dedicated LiPo charger that handles the full CC/CV cycle and termination, costs a couple of dollars, and takes `5V` in over USB. It sits between your USB port and the cell. Runtime depends on the battery's capacity and the amplifier's draw; the Bluetooth chips themselves sip power, so the amp and the cell size set how long it plays.

Sizing the battery

The math is worth running before you commit to a cup. A Bluetooth audio module pulls roughly `15-30mA` at normal volume, and a small amp adds another `~20-50mA`, so total system draw at normal listening lands around `40-80mA`. A `500mAh` LiPo then gives you roughly `6-12 hours` of playback, fine for a DIY build. Commercial headphones target `20-30 hours`, but that needs a `1500-2500mAh` cell, and those are large and heavy. For a build that actually fits in a cup, a small cell and shorter runtime is the honest tradeoff.

If the module doesn't have a built-in amp, the common DIY choice is the `PAM8403`, an inexpensive Class-D amp board that works well for headphone duty. That's the amp draw the sizing math above accounts for.

Then there's the charging port to place on the cup, the antenna to position where it isn't buried in metal or shielding that would kill its range, and the controls, power, pairing, play and pause, volume, to wire to the board. None of it is hard individually, but together it's a real subsystem that changes the weight, the complexity, and the maintenance story of your headphone, because a battery is a finite-life part that will eventually need replacing.

Two more realities worth holding onto. Wireless audio is compressed, even "lossless" Bluetooth works within constraints a wire doesn't, so a cable still wins for ultimate fidelity, and many builders wire the board so the headphone also works passively through a cable, giving you wireless convenience and a wired-fidelity option in the same set. And Bluetooth has latency, which matters for video and gaming; low-latency modes and LE Audio help, but they don't make it disappear.

> TIP
>
> Choose your module by matching its codec support to the devices you actually use. An LDAC or aptX board buys you nothing if your phone doesn't speak that codec, because it'll just fall back to SBC anyway. Check what your phone and your other sources support first, then pick the board to match, not the other way around, which is how people end up paying for codecs they can never use.

Common Mistakes

⚠ CAUTION

- **Buying for a codec your source can't use.** A high-end-codec board falls back to SBC if the phone doesn't share that codec. Match the board to your devices.
- **Treating Bluetooth as a small add-on.** It's a subsystem: board, battery, charging, antenna, controls. Plan it as one.
- **Burying the antenna.** Metal and shielding around the antenna wreck range. Give it a clear spot.
- **Forgetting the battery is a finite-life part.** It will age and need replacing. Design so it can be serviced.
- **Expecting wireless to match wired fidelity.** It's compressed audio. Wire the board for passive cable use too, and keep a wire for critical listening.

What's Next

Wireless and noise cancelling are close cousins, the same builds that go cordless often want to go quiet, and they share the same hard questions about power and electronics. The next chapter takes an honest look at what active noise cancelling really demands: active noise cancelling (Chapter 31).

CHAPTER 31

Active Noise Cancelling

An honest look at how active noise cancelling works, why it's one of the hardest things to DIY, and what's actually realistic for a builder.

11 min read   makerphones.com/learn/active-noise-cancelling

Active noise cancelling is near the top of every "can I add this?" list, and it's also one of the hardest things to do well in a DIY headphone. This chapter is an honest look at how it works, why it's so difficult, and what's genuinely realistic for a builder, so you don't burn a month chasing something that was never going to work. The short version: nail your passive isolation first, and treat ANC as an advanced sub-project only if the engineering challenge is itself what you want.

Passive versus active

Start by separating two different things. Passive isolation is what a good seal and a closed back give you for free, and it works well at high frequencies, blocking the hiss, voices, and clatter around you. What passive isolation does poorly is the low-frequency stuff: the engine drone, the rumble of a train or plane, the HVAC hum. Those long wavelengths leak through almost any passive barrier.

Active noise cancelling exists to attack exactly that low-frequency rumble that passive blocking can't touch. See open vs closed back design (Chapter 4) for the passive half of the picture.

The numbers make the division of labor concrete. A well-sealed closed-back gives you roughly `15-25dB` of passive isolation across the range, and commercial ANC adds another `~20-30dB` on top, but only down low, below `~300Hz`, where passive blocking is weakest. So ANC is for the low frequencies, and a good seal does the rest. If you're trying to kill aircraft drone, subway rumble, or HVAC hum, that low-end help is exactly what you want; if you're mostly dealing with office chatter and traffic up higher, a good seal plus a bit of music gets you most of the way without any of the ANC headache.

How ANC works

Here's the basic trick: a microphone samples the ambient noise, the electronics generate an inverted, anti-phase copy of it, the driver plays that anti-noise, and the original sound and its inverse cancel at your ear. It's destructive interference, applied in real time.

Cancellation works best at low frequencies, where the wavelengths are long and the timing is forgiving. As frequency climbs, the wavelengths shrink, the timing and phase have to be exact to a fraction of a cycle, and cancellation gets harder and then effectively impossible. This is why ANC handles low rumble well and leaves the higher-frequency blocking to the passive seal.

The three topologies

Feedforward. The mic sits on the outside of the cup, samples the noise before it reaches your ear, and the electronics play the predicted inverse. It can work over a wider frequency range, but it's sensitive to fit and to wind blowing across the external mic. There's a ceiling, though: the distance from the external mic to the driver sets how high it can cancel, and for practical over-ear dimensions feedforward works well up to `~500-1000Hz` and becomes progressively less effective above that. Anything higher is the passive seal's job.

Feedback. The mic sits inside the cup, near your ear, and measures the actual error, what you're really hearing, then corrects it. It's more forgiving of fit and strong at low frequencies, but it has a limited bandwidth and can become unstable if the loop isn't carefully tuned.

Hybrid. Both mics together, combining the strengths of each. This is what most good commercial ANC uses, and it's also the most complex to design and tune.

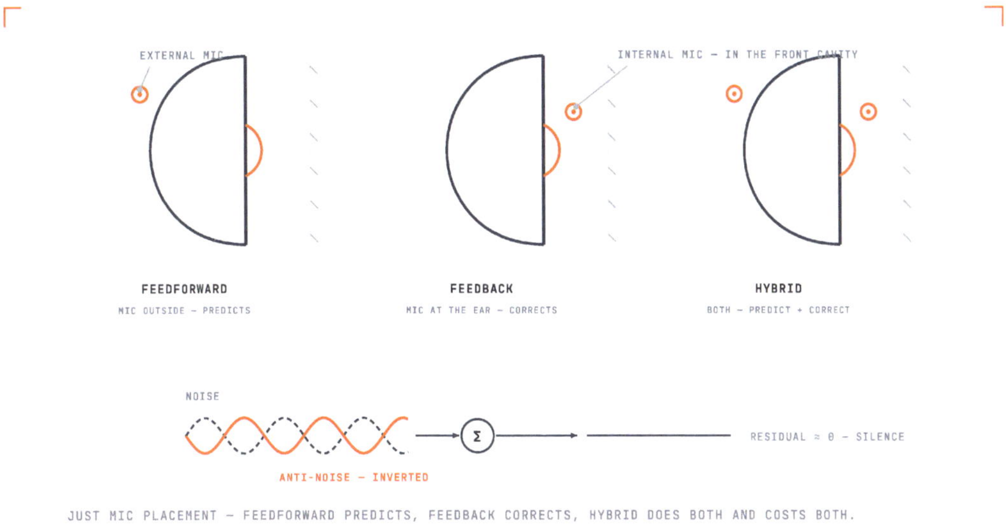

Fig. 31.1 – The three ANC topologies are just mic placement, feedforward predicts, feedback corrects, hybrid does both, and underneath, the anti-noise summing that makes it work.

Why this is brutal for DIY

ANC is not a module you bolt on. It's a real-time control loop wrapped tightly around the specific acoustics of your cup and your fit, and that's where the difficulty lives.

The anti-noise has to arrive at the right place, at the right time, with the right amplitude and the right phase, through the acoustics of *your* particular cup, with *your* particular seal. Small errors don't just reduce the effect; they can make the noise worse, or send the loop into oscillation. A feedback loop that isn't carefully stabilized squeals. That howl you've heard from a misbehaving system is a control loop gone unstable.

And the acoustics are coupled to the mechanical design at every turn: mic placement, cup volume, driver behavior, and seal all interact. You can't tune the electronics in isolation, because the plant they're controlling is your headphone. Commercial ANC is the product of dedicated DSP chips, proprietary tuning, and a great deal of control-systems engineering. The chips are available, but tuning them to a one-off custom cup is the hard ninety percent, and there's no shortcut around it.

What's realistic

Honestly: full custom ANC built from scratch is beyond most DIY, and even when it's possible the effort-to-result ratio is poor compared to almost anything else you could spend that time on. Here are the realistic paths.

Start with excellent passive isolation. A great seal and a well-damped closed cup block more high-frequency noise than people expect, no electronics involved, and that's most of what people actually want from "noise cancelling" on a commute. Put your design energy here; it's the highest-return thing you can do.

If you specifically want ANC, integrating a pre-engineered ANC module designed for headphones is far more tractable than rolling your own DSP from nothing, but you still have to fit it to your acoustics and live within the module's assumptions about cup volume and mic placement, so expect to tinker. Call it a weekend of work, with results that vary wildly from module to module.

If you actually want to play with the algorithms rather than just bolt on someone else's, a programmable audio DSP platform is the sweet spot. The `Teensy 4.0` with its audio library, or the `Daisy Seed` audio platform, can both run ANC code, and the latency can be tuned down to acceptable levels with careful firmware work. This is the most practical way to experiment with ANC in a DIY build, more involved than a pre-built module but nowhere near a from-scratch DSP. Budget several weeks if you're comfortable with embedded programming; the Teensy audio library has community ANC examples to start from. Designing your own ANC algorithm from nothing, by contrast, is graduate-level DSP work and not a realistic DIY project for most builders.

And treat ANC as the kind of advanced, open-ended project you take on because chasing loop stability for its own sake actually appeals to you.

> TIP
>
> Before you go anywhere near ANC, max out passive isolation. Most of what people want from "noise cancelling" on a commute, a proper seal and a well-damped closed cup already deliver. ANC is for the low rumble that's left, and it's worth attempting only once the passive side is as good as you can make it.

Common Mistakes

⚠ CAUTION

- **Expecting to bolt ANC on as a module.** It's co-designed with the acoustics: mic placement, cup volume, and seal all interact. There's no drop-in.
- **Underestimating loop stability.** A feedback loop that isn't carefully tuned oscillates and squeals. Stability is real control-systems work.
- **Skipping passive isolation.** Trying to cancel everything actively is the hard way to do what a good seal does easily up high. Do the passive part first.
- **Chasing ANC before the build is solid.** It's the last thing to attempt, not the first. Get the headphone itself right.
- **Expecting DIY ANC to match commercial.** It generally won't. Commercial ANC represents enormous dedicated engineering. Calibrate your expectations.

What's Next

ANC and voice features often travel together, and the mics ANC uses are cousins of the ones you'd add for calls. The next special topic is the much more approachable business of putting a microphone in your build: microphone integration (Chapter 32).

CHAPTER 32

Microphone Integration

Adding a mic to turn your headphones into a headset: mic types, boom versus inline placement, and the wiring and connector choices that actually matter.

11 min read makerphones.com/learn/microphone-integration

Adding a microphone turns a pair of headphones into a headset, for calls, gaming, streaming, or anything voice. Compared to active noise cancelling, this is refreshingly approachable: no real-time control loops, no co-designed acoustics, just a sensible set of choices about what mic to use, where to put it, and how to wire it. But those choices are real, and they affect both voice quality and whether the thing works at all on the device you plug it into.

Mic types

Electret condenser. Cheap, common, and good enough for most voice work, the electret is the workhorse of headset boom mics. It needs a small bias voltage, “plug-in power”, supplied by the host device, which most headset jacks and audio interfaces provide.

MEMS. Tiny silicon microphones, the kind in phones and laptops. They come in analog and digital flavors and are ideal where space is tight, like an inline or in-cup mic. Compact and consistent, they’re increasingly the default for small integrations.

Dynamic. More rugged and a favorite for broadcast and streaming desk mics, dynamic elements have lower output and are less common on headset booms, but they shrug off handling and plosives well.

Placement and form

Placement matters more for intelligibility than the element itself, getting the mic close to your mouth does more for clean voice than any upgrade in the capsule.

Boom mic. An arm carries the mic near your mouth. This is the best choice for voice quality and noise rejection, and it's what gaming and broadcast headsets use. A boom can be fixed, detachable, or flip-to-mute, and a directional or noise-cancelling element near the mouth rejects room noise nicely.

Inline mic. On the cable, like phone earbuds. Convenient, but the mic is far from your mouth and picks up more of the room, so voice quality suffers. Fine for casual calls, not for streaming.

In-cup mic. A mic built into the earcup is the simplest to wire and the worst for voice, because it's the farthest from your mouth. Reserve it for cases where convenience beats clarity.

Wiring and connectors

Most headset problems actually live here, so it's worth getting right.

A stereo headphone uses three conductors: tip, ring, sleeve (TRS): left, right, and ground. Add a microphone and you need a fourth conductor, which gives you TRRS: left, right, ground, and mic. TRRS on a single 3.5 mm plug is the headset standard on phones and laptops.

There are two competing TRRS pinouts, and they are not interchangeable. CTIA is the modern standard used by essentially all current phones; OMTP is the older arrangement, with the mic and ground positions swapped. Plug a headset wired for one standard into a device expecting the other and the mic won't work, or the audio comes out wrong. If a headset misbehaves on a particular device, a pinout mismatch is the first thing to check.

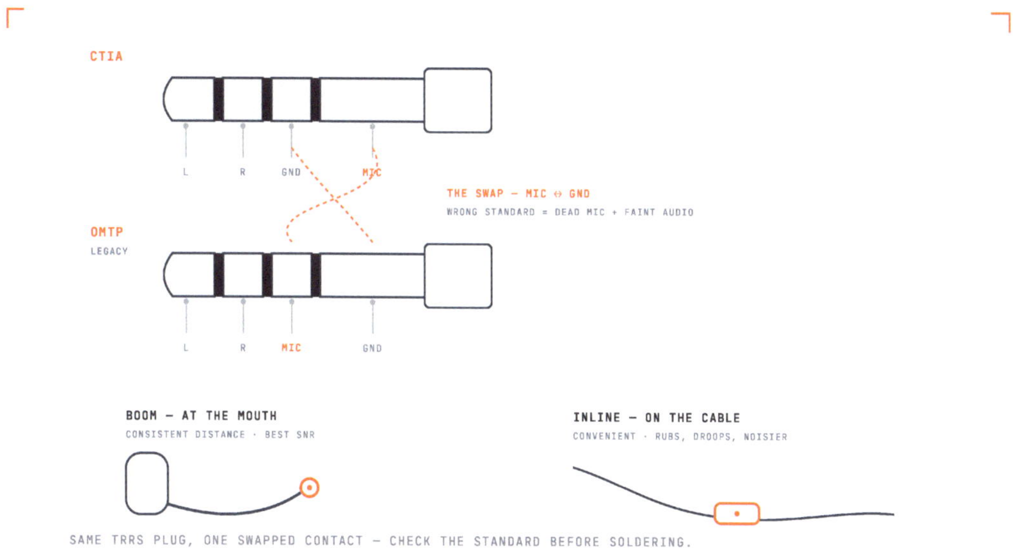

Fig. 32.1 — CTIA vs OMTP on the same TRRS plug, the two standards swap mic and ground, and where the capsule lives: boom at the mouth or inline on the cable.

The alternatives: on PCs with separate jacks, you can split into two plugs: a 3.5 mm headphone plug and a separate mic plug. Or you can go USB, where the mic and audio are digitized on board, which needs an onboard DAC and ADC and is a more involved build, but sidesteps the analog pinout question entirely.

Inside the build, route the mic wire with care. Use shielded mic cable to avoid picking up hum, keep it away from the driver leads, and give it strain relief at the connector and the boom joint, the same discipline from cables, connectors, and hardware (Chapter 12) and driver mounting and assembly (Chapter 16). Mic signals are low-level and far more vulnerable to noise than your driver leads.

A few practical notes

A small connector on the cup that makes the boom mic detachable is the most flexible arrangement, headset when you want it, clean headphones when you don't. Remember that an electret needs the host to supply plug-in power; most jacks and interfaces do, but a few don't, and a mic that's dead on one device but fine on another may just be missing its bias. Sidetone, hearing your own voice in the headphones, is usually handled by the host or its software rather than by the headphone itself, but it's worth knowing the term when someone asks why they can't hear themselves. And always test on the actual target device, because pinout and power vary between a phone, a PC, and an audio interface.

Getting clean voice

A few habits do more for how you sound than any capsule upgrade. Position a boom mic just off to the side of your mouth rather than dead in front of it, directly in the breath stream is where plosive pops and breath noise come from. A cheap foam windscreen over the element kills most of those pops and breath sounds for a dollar. Set the gain on the host so your normal speaking voice sits well below clipping but comfortably above the noise floor; too hot and you clip and distort, too low and you bury the voice in hiss. And remember that a directional element placed close to your mouth beats any amount of software noise suppression. The reliable way to judge any of this is to record yourself and listen back, rather than trusting "can you hear me okay?"

> TIP
>
> Make the boom mic detachable with a small connector on the cup. You get a proper near-mouth mic for calls and gaming when you want it, and clean, mic-free headphones when you don't, and it's easy to wire. A flip-to-mute or detachable boom also keeps the mic out of your face when you're just listening.

Common Mistakes

⚠ CAUTION

- **TRRS pinout mismatch.** CTIA versus OMTP will leave the mic dead or the audio wrong. Match the device's standard: CTIA for modern phones.
- **Unshielded mic wire next to driver leads.** Low-level mic signals pick up hum easily. Use shielded cable and route it away from the driver wiring.
- **Inline placement when voice quality matters.** Far from the mouth means more room noise. Use a boom for streaming or gaming.
- **Forgetting plug-in power.** An electret needs bias from the host. A mic that's dead on one device but works on another is often just missing its power.
- **No strain relief on the mic wire.** The boom joint and the connector flex. Anchor them, or the thin mic wire fails first.

What's Next

The last special topic is a different world entirely: the headphone shrinks until it lives inside your ear canal, the seal becomes everything, and the craft turns to tiny drivers, tuned tubes, and shells molded to your ear: custom IEM design (Chapter 33).

CHAPTER 33

Custom IEM Design

How in-ear monitor design differs from over-ears: tiny drivers, tuned tubes, crossovers, and shells molded to your ear, and a realistic path into building them.

11 min read makerphones.com/learn/custom-iem-design

In-ear monitors are a different discipline from over-ears, but they run on the same fundamentals: the seal, the driver, the chamber, the damping, just shrunk down into your ear canal. For a maker, custom IEMs are among the most rewarding and craft-heavy builds there are, and 3D printing has put the once-exclusive custom-shell process within reach. This chapter maps how IEM design differs from everything you've learned so far, and what a realistic path into it looks like.

How IEMs differ

The biggest shift is that the "cup" is now your ear canal, and the seal is even more decisive than it is over-ears. And it was already decisive there. With an IEM a poor seal doesn't just soften the bass, it eliminates it, instantly and completely. There's no large chamber doing acoustic work; the sound is dominated by the canal, the little tube or bore that carries sound to the nozzle, and the seal. The drivers are tiny and sit very close to your eardrum, which changes everything about how you tune them.

Driver types in IEMs

Dynamic (DD). A miniature moving-coil driver, the same principle as an over-ear driver shrunk down. It gives good bass, single-driver coherence, and simplicity: one driver, no crossover.

Balanced armature (BA). Tiny, efficient, and fast, a BA covers a limited frequency band well, so multi-BA designs stack several, say a bass armature, a midrange armature, and a treble armature, and split the signal between them.

See driver technologies (Chapter 6) for how these differ from dynamics. The two major BA manufacturers are `Knowles` and `Sonion`; their datasheets are publicly available, with frequency response curves, dimensions, and suggested acoustic configurations, and both sell samples directly in small quantities through their distributor pages.

Hybrid. A dynamic driver for bass paired with balanced armatures for mids and treble, combining the DD's low-end with the BA's speed up top. This is a common arrangement in higher-end IEMs.

Planar and electrostatic IEMs exist too, but they're niche; most builds start from a dynamic or a BA arrangement.

Multi-driver and crossovers

The moment you use more than one driver, you need a crossover to send each frequency band to the right driver. And in an IEM that crossover is a handful of tiny components packed into a shell the size of the end of your finger. Tuning it is the hard electrical-and-acoustic work of multi-driver IEM design, and it's unforgiving at this scale.

Just as important, and often underestimated, are the bore acoustics. The little tubes carrying sound from each driver to the nozzle have their own resonances, and tube length, diameter, and acoustic damping, small mesh dampers placed in the bores, are primary voicing tools. This is the IEM's version of chamber design and damping rolled together: see acoustic chamber design (Chapter 15) for the same principles at full size. Many an IEM is voiced as much by its tubes and dampers as by its drivers.

Don't guess at those values when you don't have to. `Knowles` and `Sonion` publish reference tubing configurations and damper resistance values in their datasheets: genuine starting points for tube lengths and crossover behavior, where a given damper resistance acts as an acoustic low-pass and sets where the driver rolls off. Starting from the manufacturer's published numbers is far more efficient than designing the bores from first principles, and guessing instead is one of the most common ways a multi-driver build goes sideways.

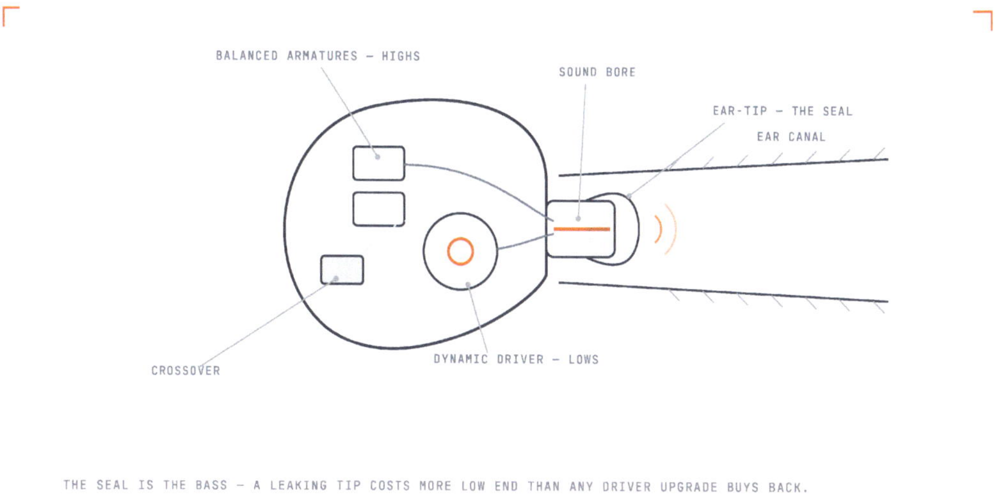

Fig. 33.1 – Inside a multi-driver IEM: balanced armatures and a dynamic driver, the crossover, sound bores to the nozzle, and the ear-tip seal in the canal.

The shell and the fit

The shell is where most of the maker craft lives, and where you pick your route: universal or custom.

A universal-fit IEM uses a generic shell and silicone or foam tips to seal against your canal. It's the easier route, the seal is good but not perfect, and the tips do the sealing work.

A custom-fit IEM is molded to your ear from an impression, giving a perfect seal, all-day comfort, and excellent isolation. And it's the craft-heavy path. The process starts with an ear impression, taken by an audiologist who fills your ear and canal with a silicone compound, then scanning or casting that impression and building the shell, 3D printed in resin, or hand-built in acrylic, around the drivers and tubes. The seal and fit dominate the result, and 3D printing from a scanned impression is what has opened that door to makers.

For the printing itself, use resin (SLA/DLP), not FDM: it produces the fine detail and smooth inner surfaces these tiny shells need, and FDM's layer lines and rougher walls change the acoustic behavior inside such a small volume. The good news is the hardware is cheap: printers like the `Elegoo Mars` or `Anycubic Photon` run roughly `$200-300` and produce IEM-quality shells.

FDM is fine for checking the rough shape, but print anything you actually intend to wear in resin.

Assembly realities

Everything here is tiny and tight. Fitting drivers, tubes, a crossover, and wiring into a shell molded to the contours of your ear is fiddly, precise work, and sealing, driver placement, and tube routing are all unforgiving at this scale. The cable typically attaches with a small detachable connector, the common two-pin and MMCX standards, so a worn or damaged cable can be replaced without rebuilding the monitor. Two-pin uses `0.78mm` spacing and is the more secure of the two; MMCX rotates freely, which is handy, but it's more prone to connector wear over many cycles. Pick one standard and stick with it across the build. The cables aren't interchangeable between the two.

A realistic path

Start simple. A single dynamic-driver universal IEM is a genuinely approachable first build: one driver, no crossover, off-the-shelf tips for the seal. It teaches you the thing that matters most, that in an IEM the canal and the seal are everything, without the crossover-and-tube complexity, and it's satisfying on its own terms.

Multi-driver custom monitors are the advanced end: crossover tuning, bore acoustics, and a custom shell built from ear impressions add up to a real project. But it's an achievable one with patience and a resin printer, and it's one of the most impressive things a headphone maker can build.

> **TIP**
>
> Build a single-dynamic-driver universal IEM first. One driver, no crossover, off-the-shelf tips for the seal. It teaches you the IEM fundamentals, where the canal and the seal decide everything, without the crossover-and-tube complexity to wrestle at the same time. It's a real, satisfying build on its own, and the perfect on-ramp before you attempt a multi-driver custom.

Common Mistakes

⚠ CAUTION

- **Underestimating the seal.** In an IEM a poor seal means no bass at all, instantly. Fit comes first, before any tuning.
- **Starting with multi-BA and a crossover.** That's the deep end. Begin with a single dynamic driver and add complexity later.
- **Treating the tubes as plumbing.** Bore length, diameter, and dampers are primary voicing tools. Tune them deliberately.
- **Cramming components without a plan.** The shell is tiny. Plan driver placement, tube routing, and wiring before you assemble.
- **Skimping on the ear impression for a custom.** The perfect fit is the whole point of going custom. A bad impression wastes everything downstream.

What's Next

That's the last of the special topics, and it closes the reference side of the manual. From here the path turns hands-on: the build guides, where all of this technique and judgment finally meets a workbench, starting with a complete first build you can follow end to end. Your first single-driver build will teach you more about seal than any chapter can. Start there.

REFERENCE

Build Guides

Choosing a 3D Printer for Headphones

Which 3D printer should you buy to build headphones? A point-in-time buying guide: what actually matters for small headphone parts, what a better machine buys you, what it doesn't, and concrete picks with honest tradeoffs.

The rest of the manual is pretty thorough about *what* to print and *how*: which filament, what wall thickness, where to add clearance. What it never says is which machine to actually buy. That's the gap this page fills. And I want to be upfront about its nature: a printer recommendation is a perishable thing. Models get superseded, prices swing on sales, and a great-value pick this quarter is mid-pack the next. So treat everything below as a dated snapshot, current **as of June 2026**, not gospel. If you're reading this much later and the landscape has clearly moved on, use the feedback form at the foot of the page to flag it. This page is meant to be kept current by builders plus periodic refreshes, not frozen.

ⓘ WHERE I'M WRITING FROM

I'm restarting my current build on an Elegoo Centauri Carbon. I don't yet have long-run, hands-on mileage on it, so the picks here are reasoned current guidance from spec, reputation, and the broader community's experience, not lived test results. A hands-on follow-up will come once I've actually got real print hours on the machine.

The good news: headphone parts ask very little of a printer

Before you spend a dollar, understand what you're printing. The parts that make a headphone yours, the cups, baffle, yoke, and slider, are *small*. None of them comes close to filling a budget printer's bed. So build volume, the number most printer marketing leads with, is a non-issue here. Every machine worth buying has room to spare.

The parts are also spec'd in forgiving materials. PETG for durable, structural parts; PLA+ for quick prototypes. Both print fine on an open-bed machine with no enclosure. You do not *need* a fancy printer to build a good headphone. Plenty of excellent builds come off the cheapest bed-slingers. If you already own one, the honest answer is: start with what you have, and come back to this page when you've found its limits.

That's the baseline. The interesting question is what stepping up actually buys you.

What a better printer genuinely buys you

There are three real gains, and they map cleanly onto what builders care about.

Dimensional repeatability. This is the big one. Headphone parts have to *mate*: a baffle press-fits a driver, a slider rides in a track, two cups need to come out identical. A rigid, enclosed CoreXY machine holds its dimensions run to run far better than a flexing, open bed-slinger whose moving bed introduces tiny inconsistencies. If you've ever printed the same part twice and had one fit and one not, this is why. Consistency is the single strongest reason to step up. It's also the foundation of everything in manufacturing for consistency (Chapter 28).

Surface finish at speed. A rigid frame, input shaping, and good part cooling let a printer run fast *and* clean. You get fewer ringing artifacts, crisper overhangs, and fewer support scars on the cosmetic faces of a cup. On an open-bed machine you can get there too, but you'll be printing slower and fighting more.

Material headroom. An enclosed, heated chamber lets you reliably run ABS, ASA, and even PC: tougher, more heat-tolerant materials with a more premium hand-feel, and an easier time with heat-set inserts. An open-bed machine mostly keeps you in PLA and PETG, which, to be clear, is genuinely enough for the build. The chamber is a nice-to-have, not a requirement. The material tradeoffs themselves are covered in acoustic chambers and enclosures (Chapter 9).

What a better printer does *not* fix

This trips people up, so I'll be blunt about it.

Easy part removal is not a price-tag thing. It's a flex-plate thing, and every modern machine, $200 or $1,600, ships with a PEI-coated spring-steel sheet. You pop parts off by flexing the plate. A pricier printer does not release parts any better. Don't pay up expecting magic here; you already have the magic.

Easy cleanup is mostly a design and slicer problem, not a hardware one. How much support you have to pick off, and how cleanly it comes away, comes down to part orientation, support strategy, and whether you chamfered your overhangs in CAD. That stays in your hands no matter what you buy. 3D design for headphones (Chapter 14) is where you actually win this fight.

Multicolor and automatic material switching are irrelevant. Headphone functional parts are single-color. The multi-material systems (and the combos sold around them) add cost, complexity, and filament waste for exactly zero benefit on a cup or a yoke. If a printer's headline feature is auto color-changing, you're paying for something this hobby will never use. Skip it.

The picks (as of June 2026)

Prices are approximate and move constantly, especially on sale, so confirm before you buy. These are reasoned recommendations, ordered roughly by spend.

Elegoo Neptune 4 Pro — the "get printing today" pick (~entry-level). An open-bed bed-slinger, about the cheapest serious machine you can buy, and representative of what a lot of readers already own. It prints the whole build fine in PETG. What you give up is easy ABS/ASA and the warm-chamber ease for heat-set inserts. If your goal is to start building this weekend without overthinking it, this is the honest match.

Elegoo Centauri Carbon — the value step-up (~$300, sometimes ~$285 on sale). An enclosed CoreXY for roughly the price of a good bed-slinger: fast, rigid, and ABS/ASA-capable. It's been superseded by the Centauri Carbon 2, but the original is still a strong machine and a lot of printer for the money. This is my own restart pick. One caveat I'd press hard on: **skip the CC2 Combo.** The multicolor CANVAS system adds nothing for headphone parts, and there's a filament-cutter design flaw (Error 1231, "Filament Cut Failed") hitting a growing number of users. The single-material machine sidesteps the whole problem.

Bambu Lab P1S — the least-fuss pick (~$500). An enclosed CoreXY that is, by reputation, the most dialed-in and lowest-hassle machine in this range, with the best out-of-box surface finish. The tradeoff is real and worth naming: the software leans on the cloud and the ecosystem is fairly walled-garden. For an open-source project, that's a genuine values tension. If you mostly want parts to just *work* and that tension doesn't bother you, it's hard to beat.

Prusa Core One — the open-ethos pick (~$1,099 kit / ~$1,599 assembled). An *actively heated* chamber, the quietest of the bunch, a load-cell first layer, and the openness that matters to a project like this: published CAD, open firmware, any slicer, fully offline, lifetime support. If hardware openness is a value you hold, this is the match. Honest caveat, and reviewers say the same: it's overkill if you'll only ever print PLA and PETG. It's worth the premium only if you want real engineering materials and an open machine.

Advanced / optional — a hybrid FDM + resin (SLA) workflow. Resin gives the best surface and the most precise, isotropic parts, which makes it tempting for the cosmetic show-surface of a cup. The cost is brittleness and a real cleanup-and-post-cure tax (gloves, alcohol, UV curing, messy consumables). Treat this as a "once you're comfortable" second machine, not a first printer.

A short buying checklist

- **Build volume** — a non-issue for headphone parts. Don't pay for it.
- **Enclosed chamber** — only worth it if you actually want ABS/ASA/PC. For PLA/PETG it's idle weight.
- **Motion system / dimensional consistency** — the thing actually worth spending on. Rigid CoreXY over flexing bed-slinger if mating fit matters to you.
- **Auto bed leveling + forgiving stock profiles** — buys low-fuss reliability, which is worth real money when you're starting out.
- **Flex plate** — universal. Every modern machine has one. Don't overpay for "easy removal."
- **Multicolor** — ignore it for this hobby.

Don't want to buy at all?

You don't have to. If you'd rather not commit to a machine yet, online FDM print services and local makerspaces will print your part set from a model, which is a low-commitment way to start. That route is covered in the supplier directory, and the Daily Driver design spec has a section on getting its parts printed without your own printer. Printing a build or two through a service is a perfectly sane way to learn whether you want a printer before you own one.

Common Mistakes

> ⚠ CAUTION
>
> - **Buying for build volume you'll never use.** Headphone parts are small. A bigger bed is wasted money here.
> - **Expecting a pricier printer to fix part removal or cleanup.** It won't. Removal is the flex plate (everyone has one); cleanup is design and slicer strategy.
> - **Paying for an enclosure you won't use.** If you only print PLA and PETG, the chamber is dead weight you bought.
> - **Buying a multicolor combo for single-color functional parts.** You're paying for waste and complexity the build never touches.
> - **Forgetting that mating fit is per-printer calibration.** Don't assume someone else's clearances will fit off your machine. Print test fits and tune the numbers to *your* printer.
> - **Using plain brittle PLA on stressed parts.** The yoke and headband flex and fatigue. Use PETG (or a tougher material) for anything load-bearing; save plain PLA for prototypes.

What's Next

A printer is only half the picture; the other half is the file you feed it. 3D design for headphones `(Chapter 14)` covers modeling parts that actually print and fit, including the clearances you'll calibrate to whichever machine you land on, and sourcing components `(Chapter 8)` covers filament alongside everything else the build needs. When you're ready to point a machine at a real project, the Daily Driver design spec is the build I'm running mine against. And again: if the printer landscape has moved on since June 2026, the feedback form below is the way to keep this page honest.

Your First Build

A complete first headphone build, start to finish: a simple single-driver over-ear you can print, wire, and tune by ear, pulling the whole manual together into one project.

You've read the manual, or skipped straight here, no judgment. Either way, it's time to build your first headphone. We're keeping it simple on purpose. A first build isn't about a perfect headphone; it's about going through the whole process once. Make the beginner mistakes where they're cheap, and come out understanding how all the pieces fit together. You'll build a better one next time. Everyone does. The first build is the hardest precisely because everything is new.

We'll build the most approachable thing there is: a simple over-ear with a single dynamic driver in a 3D-printed cup. One driver, no crossover, a design you can print and tune by ear. It touches every fundamental without the hard-mode complications.

What you'll need

Materials: a single full-range dynamic driver, an inexpensive, well-regarded model in the common 40 mm class is a perfect place to start; filament (PLA or PETG) for the cups and baffles; a set of ear pads; a length of cable and a connector; some felt or fiber fill for damping; a headband, either printed or repurposed from an old pair; and basic hardware, screws or a gasket to hold the driver.

Tools: access to a 3D printer; a soldering iron and solder; a hobby knife and basic hand tools; optionally a multimeter; and, if you catch the bug, a measurement rig later on.

Buy your drivers as a matched pair, and where possible from the same batch: cheap drivers vary, and matched channels start at the source.

Stage 1 — Pick your driver

The driver is the heart of the build, so start here. For a first build, don't chase anything exotic. Pick an inexpensive dynamic driver that other builders have used successfully, so you're building on known ground rather than gambling. Driver selection guide (Chapter 7) covers what to look for, and sourcing components (Chapter 8) covers where to find it.

Stage 2 — Get your cup

You can design your own cup from scratch, 3D design for headphones (Chapter 14) walks through that, but for a first build the smarter move is to start from a known, simple design sized for your driver and modify it later once you understand what each change does. Lean toward an open or lightly vented cup. Open-back is the more forgiving choice for a beginner, because the seal is far less make-or-break than it is on a closed design. If you want to understand the volume and venting you're printing, acoustic chamber design (Chapter 15) explains the levers.

Stage 3 — Print the parts

Print the cups, baffles, and any structural pieces. Print both cups with the same settings and the same material so the pair matches: this is your first taste of manufacturing for consistency (Chapter 28). When they come off the bed, check that the driver actually seats in the bore, and clean it up with a hobby knife if it's tight. A driver forced into a too-small bore sits stressed and crooked; one that rattles won't seal.

Stage 4 — Mount and wire the driver

This is the stage where seals are won or lost, so slow down here. Seat the driver against a soft gasket and hold it with screws or a clamp, don't glue it, because you'll want it back out to retune. Solder your leads and get the polarity right: mark positive and keep both channels in phase, or the bass cancels and the stereo image collapses into a vague mess. Add strain relief where the wire meets the driver and where the cable enters the cup.

Before you close anything up, wire it and play both channels with a mono track. It should image as a single point dead between your ears. If it sounds wide or hollow, one channel is out of phase: fix it now, not after you've sealed the cup. The full detail lives in driver mounting and assembly (Chapter 16) and cables, connectors, and hardware (Chapter 12).

Stage 5 — Fit the pads

Pads aren't only about comfort: they set the front cavity and the seal, so they genuinely shape the sound. Fit them evenly on both sides, and treat them as part of the design rather than an afterthought. Ear pads and comfort (Chapter 10) covers the acoustic side of the choice.

Stage 6 — First listen, then tune by ear

Plug in and listen. It probably won't sound finished, and that is completely normal, you've built a raw instrument, and now you voice it. Tune by ear, one change at a time: add a little felt to the rear chamber, listen, decide on the next move. Damping strategy and application (Chapter 17) tells you where damping goes and what each placement fixes, and warns you, rightly, against the beginner's urge to stuff the cup full. This is the part that hooks people: you move a scrap of felt and you can hear it. One change at a time, and the headphone starts sounding like yours.

Stage 7 — (Optional) Measure

You can build a genuinely good first headphone entirely by ear, and plenty of people do. But the moment you want to match your channels precisely or actually see what your tuning is doing, a measurement rig pays off. Why measure headphones (Chapter 18) makes the case, and budget measurement setup (Chapter 19) shows how to put a usable rig together cheaply, or even print part of it yourself.

Common first-build mistakes

- **Reversed polarity on one channel:** thin sound, no bass, vague image. Check with a mono track before you seal up.
- **Gluing the driver in:** you'll want it out to retune. Use a gasket and clamp.
- **Over-damping:** stuffing the cup kills the life. Less is more; add in small steps.
- **Mismatched cups:** build both the same way, in the same session, with the same materials.
- **Chasing perfection:** it's a first build. Finish it, learn from it, and put what you learned into the next one.

Where to go next

Build a second one. Apply what you learned, and notice how much faster and more confident the second build feels: that's the whole point of the first. Once you trust your seals, try a closed design. Get a measurement rig and start tuning by data. Share your build with the community and see what others did differently. The fundamentals don't change; what changes is you, getting better at applying them. The rest is reps.

Simple Open-Back Build

A clean, intentional open-back build, leaning into the airy, out-of-your-head sound open-back does best. The forgiving design, done with enough care to actually sound open.

Open-back is where a lot of builders fall in love with this hobby. It's the more forgiving design, the rear wave escapes into the room instead of fighting you inside a sealed cup, and it rewards you with the out-of-your-head sound that open-backs are known for. If your first build was about learning the process end to end, this one is about doing it with intent: building a clean open-back that actually sounds open.

The trade-off, stated plainly up front: open-back leaks both ways. It won't isolate you from the room or the room from anyone near you, and it gives up some deep-bass slam compared to a sealed design. In exchange you get soundstage, naturalness, and an easier build. For listening at home in a quiet space, it's a wonderful place to live.

What you'll need

Materials: a dynamic driver suited to an open design; filament (PLA is fine here); an open or grille-style rear, a printed grille, a metal mesh, or an open frame; a set of airy ear pads; cable and a connector; a little felt or fiber for light damping; a headband; and basic hardware and a gasket.

Tools: a 3D printer; a soldering iron and solder; hand tools; optionally a multimeter; and a measurement rig if you want to verify your work.

Stage 1 — Pick a driver suited to open-back

Because an open-back gives up some sub-bass by nature, a driver with a present, articulate mid-bass and a smooth top end suits it well: you're playing to the design's strengths rather than fighting its weaknesses. Driver selection guide `(Chapter 7)` covers what to look for; buy a matched pair.

Stage 2 — Design the cup with an open back

The defining feature here is the open rear that lets the back wave radiate freely. Print a grille, fit a mesh, or leave an open frame: the goal is a large, unobstructed rear opening. Keep the front cavity, between driver and pad, sensible and consistent. If you want to understand *why* the open back changes the sound so much, open vs closed-back design `(Chapter 4)` lays out the acoustics, and 3D design for headphones `(Chapter 14)` covers the modeling.

Stage 3 — Print the parts

Print both cups with the same settings and material so the pair matches: the consistency habit from manufacturing for consistency `(Chapter 28)` applies to every build. Check the driver seats cleanly in the baffle.

Stage 4 — Mount and wire

Open at the back doesn't mean sloppy at the front: the driver-to-baffle seal still matters, because the front cavity is what loads the driver into your ear. Gasket the driver, get the polarity right and both channels in phase, add strain relief, and play a mono track to confirm the image sits dead center before you finish up. Full detail in driver mounting and assembly `(Chapter 16)` and cables, connectors, and hardware `(Chapter 12)`.

Stage 5 — Fit the pads

Open-back designs pair beautifully with airy, breathable pads, velour is a common choice. Fit them evenly on both sides, and treat the pad as part of the acoustic design, per ear pads and comfort `(Chapter 10)`.

Stage 6 — Tune, gently

Here's the open-back lesson: you need *less* damping than a closed design, because the back wave escapes into the room instead of bouncing around a sealed cavity. Your job is mostly taming reflections off the cup interior and smoothing the driver's own peaks, not fighting trapped air. A light lining and perhaps a thin layer behind the driver is often plenty. The temptation to add more material is strong and almost always wrong here: an open-back wants to breathe. Damping strategy and application `(Chapter 17)` covers what little you do add.

Stage 7 — Listen, and verify if you like

You're chasing that open, spacious presentation, and an open-back done well throws a soundstage a sealed cup just can't. If you want to check your channel matching or see what your light tuning did, why measure headphones (Chapter 18) makes the case for a quick measurement.

Assembly, step by step

The stages above are the map; this is the turn-by-turn. None of it is hard, but a few small habits are the difference between a pair that seals and sounds right and a pair that fights you. Work in this order: most first-build mistakes happen at the driver gasket and the soldering, so slow down there.

Set the heat-set inserts

If your baffle uses heat-set inserts for the driver or baffle screws, set them with your soldering iron at `200–220°C`. Sit each insert on its hole, bring the iron tip down onto the insert itself, and let it sink under gravity plus gentle pressure, don't force it. Stop when the insert is flush or about `0.2mm` below the surface, then let it cool fully before you touch it.

> TIP
>
> No proper heat-set tip? A small bolt with the head cut off works fine: the goal is to carry heat evenly through the whole insert, not nip the top edge. A crooked insert means a crooked screw, and a crooked screw means an acoustic leak, so take your time and keep each one square.

Mount the driver

Press gasket tape into the channel around the driver opening, cut cleanly and seated firmly all the way around. This is the seal between driver and baffle, and any gap here is a leak that costs you bass: it's the front-seal point the mount-and-wire stage above keeps hammering on.

Before the driver goes in, verify polarity. Find the positive terminal on each driver, usually a red dot, a +, or a color-coded terminal, and write down in your notes which terminal is positive on *your* specific drivers. You'll want that when you wire up. Set the driver against the gasket so it sits flush or slightly recessed, never proud, then hold it with a few dots of hot glue on the driver frame itself, not on the gasket. That keeps it put while you solder but lets you pull it later.

Don't commit with permanent adhesive yet; save that for after everything tests good.

Solder and verify🔗

Strip and tin your wires *before* you ever touch the driver terminals. Those terminals are small and hate heat, so the less time the iron spends on them the better, pre-tinned wire lets you make the joint in one to two quick seconds rather than five slow ones. Let each joint cool before you move it; it should look shiny and smooth, not grey or balled up.

Then prove it with a meter before you close anything up. Set the multimeter to resistance and measure between each channel pin and ground at the plug: you should read roughly the driver's impedance, around ~32Ω for a 32Ω driver, on each channel. An open circuit means a cold joint or a broken wire. Find it now, while everything is still open and easy to reach.

> ⚠ CAUTION
>
> Don't skip the continuity check to save five minutes. Once the baffle is screwed down and the pads are on, a cold joint turns into a teardown. The meter reading is your proof the channel is actually alive.

Close up the baffle🔗

Run a thin bead of glue around the gasket channel on the cup side to reinforce the seal, not across the whole mating face, or you'll never get back in. Set the baffle on, start all the screws finger-tight, then snug them in an X pattern rather than going around the circle. You're compressing the gasket evenly; chase one screw down first and the baffle tilts and the seal goes with it.

Then flex-test it: hold the assembled cup and press lightly around the baffle edge. Any flex is a spot that isn't sealing. A very thin bead of glue along the *exterior* seam closes minor gaps, but if there's real flex, fix the root cause rather than caulking over a warped part.

Damp it — lightly, on purpose🔗

The open-back discipline from the tuning stage above becomes physical here. Start with a single layer of felt on the rear wall of the cup, cut to fit without blocking the vents, plus about a golf ball's worth of loose fiberfill in the chamber, loosely placed, not packed. That's the whole starting recipe.

Deliberately under-damp this first listen. You'll hear something a touch bass-heavy and a little resonant, and that's correct: it's a starting point. You add damping a bit at a time and listen again; you can't take it back out once it's changed the sound. Damping strategy and application (Chapter 17) walks the full iteration.

Route the cable

Give the cable a small slack loop, roughly `30mm`, at each cup end so the cup can move without ever pulling the wire taut against the driver terminals. Too little slack and every head turn tugs the joint you just soldered. Anchor the cable at each end with a small zip tie through a tie point so it stays put and doesn't foul the adjustment hardware.

First listen

Before you reach for a measurement rig, run these checks by ear with music you know well. They catch the build problems a measurement won't tell you in plain language.

1. **Mono center image.** Play a centered mono signal, a voice or a single instrument, and it should sit dead center in your head. If it pulls left or right, recheck your wiring polarity.
2. **Rattle and bottoming.** Play something bass-heavy at moderate volume and listen for buzz or rattle (usually something loose in the cup) or gritty distortion (the driver bottoming out, check the gasket isn't so thick it's pushing the cone into something).
3. **Channel balance.** Swap the cups left-to-right on your head. If the same physical cup is always louder, that's a driver mismatch or wiring issue, not your ears or the fit.
4. **Seal-break check.** Press the cups firmly to your ears and note the bass, then slowly break the seal on one side: you should hear the bass change clearly. No change means the cup isn't sealing against your head; look at pad depth and cup angle.

Then, before you start analyzing anything: enjoy them for a minute. You built these.

Common mistakes

- **Over-damping**: open-backs need very little. Stuffing the cup kills the openness you built it for.
- **Expecting closed-back bass**: open-back trades deep slam for spaciousness. You knew that going in.

- **Choking the rear opening:** too small a grille or vent makes an open-back behave half-closed and boxy. Keep it open.
- **Mismatched open area between cups:** asymmetry skews the stereo image. Build both rears identically.
- **Ignoring the front seal:** open at the back, sealed at the baffle. Don't get sloppy at the driver.

Where to go next

The natural next challenge is a closed-back build, for isolation and bass, the harder sibling to this one, where everything that makes open-back forgiving starts working against you. Or refine this open-back with measurement and small tuning passes until it's exactly where you want it.

Closed-Back Studio Build

A closed-back headphone built for isolation and controlled bass, the harder, less forgiving build, where the seal is everything and a sealed cup wants to boom and ring until you tame it.

Closed-back is the harder build, and a studio-oriented closed-back is the most demanding of the approachable projects, which is exactly why it's worth doing. Sealing the back gives you isolation, so you can monitor or track without sound leaking out or the room leaking in, and it gives you stronger, more authoritative bass. But everything that made the open-back forgiving is now working against you: the rear wave is trapped, the seal is make-or-break, and a sealed cup wants to resonate and boom if you don't control it. Get it right and you have a headphone you can actually work in.

This is a real step up, so it's worth doing after you've got a build or two behind you. The reward is a headphone built for a job, monitoring while tracking, listening in a loud room, working without disturbing anyone, where isolation, controlled bass, and a response you can trust matter more than easy fun.

What you'll need

Materials: a driver that handles a sealed load well, controlled bass rather than already-boomy; filament, with PETG worth considering here for its extra mass and damping in the shell; a sealed cup design, possibly with a small tuned vent; good isolating pads that seal well, leatherette or protein leather seal better than velour; cable and a connector; more damping material than an open-back build needs: felt, foam, fiber fill, and modeling clay for the baffle; a headband; and hardware with gaskets for a proper seal.

Tools: a 3D printer; a soldering iron and solder; hand tools; a multimeter is worth having; and a measurement rig is strongly recommended, because closed-back hides its problems and a rig is what lets you find them.

Stage 1 — Pick a driver that suits a sealed design 🔗

A closed-back reinforces bass, so a driver that's already heavy down low can turn boomy once you seal it in. A driver with a clean, controlled low end gives you something you can actually tune. Driver selection guide `(Chapter 7)` covers the choice.

Stage 2 — Design a sealed cup with the right internal volume 🔗

This is the crux of the whole build. Too small a sealed volume stiffens the trapped air, producing a one-note bass hump and a resonance you'll spend the rest of the build fighting; enough volume keeps the bass even and natural. You can also add a small, carefully tuned vent to relieve the sealed-chamber resonance, a semi-closed approach that trades a little isolation for smoother bass. Acoustic chamber design `(Chapter 15)` is essential reading here, and open vs closed-back design `(Chapter 4)` explains why the sealed cavity behaves the way it does.

Stage 3 — Print the parts 🔗

Print both cups identically. Thicker walls or a denser material like PETG help damp the shell so it doesn't ring like a drum under bass, and a quiet shell is part of a clean closed-back, as manufacturing for consistency `(Chapter 28)` and the resonance discussion both stress.

Stage 4 — Mount and wire with an airtight seal 🔗

In a closed-back, leaks are the enemy. Any leak around the driver or a poor pad seal collapses the very bass you sealed the design to get, and undoes your isolation. Gasket the driver carefully, add strain relief, get the polarity right and both channels in phase, run a mono track to check the image, and then pressure-check the seal by gently pressing the cups. If the bass changes dramatically, you have a leak to find. Driver mounting and assembly `(Chapter 16)` and cables, connectors, and hardware `(Chapter 12)` cover the details.

Stage 5 — Fit isolating pads 🔗

Sealing pads with firm, even contact do double duty here: the pad seal is half your isolation and half your bass. Leatherette and protein-leather pads seal better than open velour, at some cost in breathability. Fit them evenly and treat the seal as load-bearing, per ear pads and comfort `(Chapter 10)`.

Stage 6 — Tune with damping

Closed-back is where damping does the heavy lifting, because now you're fighting trapped-air resonances and internal reflections inside a sealed cup. Line the walls to kill reflections, add fill to tame the rear-chamber resonance and the bass hump, and use a thin layer to smooth the driver's peaks, but balance it carefully, because too little leaves boom and ring while too much deadens the life out of it. This is iterative work, one change at a time, and it's exactly where a measurement rig turns guesswork into seeing the problem. Damping strategy and application (Chapter 17) and tuning with damping (Chapter 22) walk through it.

Stage 7 — Measure

This is the build where measurement matters most. The sealed bass hump, the chamber resonance, the channel matching, and your isolation are all hard to judge by ear alone, and easy to see on a graph. If you've been building by ear so far, the closed-back studio project is the one that makes a measurement rig worth it. Why measure headphones (Chapter 18) makes the case, and budget measurement setup (Chapter 19) shows how to put one together affordably.

Common mistakes

- **Leaks:** the cardinal closed-back sin. A leak kills the bass and the isolation you sealed the cup to achieve. Pressure-check everything.
- **Sealed volume too small:** produces a one-note, boomy bass and a nasty resonance no amount of felt fully fixes. Get the volume right first.
- **Under-damping:** boom and ring from trapped air and internal reflections. Closed-back needs real damping.
- **Over-damping:** dead and lifeless. The sealed cup tempts you to overstuff; resist past the point that solves the problem.
- **A ringy shell:** thin printed walls drum under bass. Stiffen the structure; don't just line it with foam.
- **Skipping measurement:** closed-back hides its problems from the ear better than any other design. Measure it.

Where to go next

This is the most demanding of the approachable builds, so finishing one well is a real milestone. From here it's refinement: measurement-driven tuning passes, dialing the vent and damping until the bass is even and the isolation holds, and then your own designs. A closed-back studio headphone is a tool you can actually work in, built with your own hands.

Daily Driver — Design Spec

The functional design specification for the Daily Driver, Makerphones' canonical first-build open-back: system architecture, acoustic approach, component specs, parametric setup, and the open design questions.

Web-only: the rotatable 3D part previews in this spec can't print. View and spin every part at **makerphones.com/learn/daily-driver-design-spec**.

Spin it in 3D: drag to orbit, scroll to zoom (it auto-rotates on load). Both earcups + the headband: cups, baffles, fork-yokes, sliders, and the steel bow. Source files: github.com/makerphones/daily-driver. **Want every part?** Pull it apart, isolate any part, and see what to print →

Status: Design phase, CAD in progress. This document is the functional spec the CAD is built against, and the reference behind the build guide, the parts list, and the measurements.

Source: The CAD, STLs, headband DXF, and this spec are developed in the open at github.com/makerphones/daily-driver, the canonical home for the design files.

Overview

The Daily Driver is the canonical first build: a 40 mm open-back (Chapter 4) headphone that's forgiving to design and to assemble, buildable by a first-timer, and good enough to keep wearing once it's done. It's fully parametric and built to be modified: change the driver, the pad, or the head size and the model follows. A pre-sourced kit may be offered later; everything here is also exactly what you need to build it yourself. For the same architecture as a hands-on walkthrough, see the simple open-back build.

Honest sonic target: a **bright, open, detailed open-back**, strong mids and treble, modest bass. That's the nature of a small driver in an open baffle, and the design leans into it rather than chasing sub-bass the hardware can't make.

System architecture

Four printed/sourced parts per side, each interface independently iterable:

```
SPRING STEEL ARC  (sourced, laser-cut flat, formed over a jig
        |
SLIDER × 2  (printed) — adjusts length, wraps the arc, carrie
        |
YOKE × 2  (printed) — connects slider to cup, allows ±20° til
        |
CUP ASSEMBLY × 2  (printed: cup shell + separate baffle plate
```

Keeping each interface separate means any one part can be reprinted and iterated without touching the others.

Acoustic approach (open-back)

- **Open rear.** The back of the cup is an open vent pattern with no sealed chamber behind it. That's what makes the design forgiving and gives the open, airy character.
- **Front cavity = pad depth.** Driver-to-ear distance is set by pad thickness, see ear pads and comfort (Chapter 10) for why. Deeper is more comfortable and slightly warmer; shallower is more present and brighter. This is the main tuning lever, and it's why the pad choice is locked into the spec.
- **Light rear damping.** A disc of acoustic felt or thin open-cell foam across the rear vent tames cone breakup and reflections: damping materials (Chapter 11) covers the options. Start light, tune by ear and measurement (Chapter 20).
- **No sealed volume to get right**: the reason an open-back is the right first build.
- **Expected response:** forward mids, extended/airy treble, gentle rolloff in the low bass.

Design decisions

Cup shell

(Chapter 9)).

- **Open-back vent**: ring of 6–8 oval slots on the rear face, ~40% open area.
- **Pad attachment**: tension-ring lip, 61–62 mm OD to accept Brainwavz HM5 and compatible pads.
- **Walls**: 3 mm minimum shell, 4 mm at structural points (baffle flange, yoke boss).
- **Print orientation**: open face down, no supports.
- **Material**: PLA+ for prototyping, PETG for durable/kit parts.

Baffle plate

- **Separate from the cup**: 4× M3 screws into heat-set inserts; removable for driver swaps and tuning.
- **Driver cutout**: acoustic aperture only; the driver sits on a ledge (the frame doesn't pass through).
- **Gasket channel**: 2 mm wide × 1.5 mm deep around the driver cutout.
- **Insert placement**: 4 inserts, symmetric, ~70% of radius from center.
- **Cable entry**: 6–7 mm hole near one edge, toward the headband side.
- **Print orientation**: flat, no supports.

Yoke

- **Separate from the cup**: lets the cup and yoke iterate independently.
- **Own pivot spec**: a 6 mm-diameter post, deliberately *not* cross-compatible with commercial (e.g. Beyerdynamic) yokes; this is the project's own interface.
- **Pivot post**: 6 mm diameter, 8 mm tall, one flat face, M3 through-hole for a retention screw.
- **Rotation**: ±20° tilt, with hard stops in the geometry to protect the cable.
- **Single axis**: tilt only; no fold mechanism on v1.
- **Left/right**: mirror images, two files.

Slider

- **Wraps the spring steel arc** on three sides.
- **Adjustment**: friction clamp: a printed spring arm presses on the steel face.
- **Detent-ready**: the boss geometry also works with punched holes in the steel for a click-fit upgrade.
- **Carries the pivot post**: 6 mm post matching the yoke bore, M3 retention screw.

Cable

- **Dual entry**: each cup has its own exit; Y-split at chin level. No routing through the headband arc: simpler wiring, easier assembly.
- **Strain relief**: raised ring at the entry; jacket clamped before it enters the cup.

Component specifications

Driver (parametric)

Design the baffle around named parameters so a specific driver can be locked late. Confirm the real numbers with calipers and print a baffle test piece (a ~20-minute print) before committing a full cup set.

PARAMETER	STARTING VALUE	NOTES
Outer frame diameter (`driver_od`)	42 mm	38–42 mm class; **measure your actual driver**
Acoustic aperture (`driver_aperture`)	35 mm	Driver sits on the ledge inside this
Gasket ledge width	3.5 mm/side	(`driver_od` − `driver_aperture`) / 2
Driver depth	10 mm	Sets minimum cup interior depth
Cutout tolerance	+0.3 mm	Actual cut ≈ 35.3 mm for fit
Impedance target	32 Ω	Works from phones/laptops, no amp

Selecting a driver. The driver selection guide (Chapter 7) has the full method; for this design look for: 40 mm nominal, 32 Ω, neodymium magnet, a *smooth* response without aggressive peaks above ~3 kHz, and ideally a published measurement from another builder. Good options sit in the $12–18 range at Parts Express or Madisound.

Budget option, with a caveat. The Dayton CE38MB-32 (~$3.49, 40 mm frame, 32 Ω, Ø35.5 mm opening, ~5.6 mm deep) is the cheapest way in, but it has a peaky upper midrange that takes extra damping work to tame, workable, but not the easiest first-build driver. Choose it only if cost is the priority, and expect to spend more time on damping.

Ear pads

- Brainwavz HM5 velour, 90 mm oval (widely available; the design is built to this pad).
- Friction-ring attachment: pushes over the cup's pad lip.
- Confirm the HM5 inner ring diameter with calipers before finalizing the lip OD (nominal ~60 mm; cup lip OD target 61–62 mm).

Spring steel arc

PARAMETER	VALUE	NOTES
Material	1095 high-carbon spring steel	Standard headband material
Thickness	0.7 mm	0.6 mm lighter/gentler, 0.8 mm firmer
Width	10 mm	Standard headband width
Target inside arc radius	~100 mm	Needs prototype iteration (see Open questions)
Developed (flat) length	~370 mm	~100 mm radius × ~90° arc + 30 mm terminations each end
Detent holes (optional)	3 mm dia., 6 mm spacing	Punched at cut time for click-adjust

Sourcing the steel. Draw the arc as a *flat developed profile* (imagine the arc unrolled), export a DXF, and order it laser-cut in 1095 spring steel from a service like SendCutSend (~$10–15 delivered at low quantity). They ship flat precision blanks; you form the curve cold over a printed radius jig. No metal shop, no minimum orders. (Builders on a tighter budget can salvage spring steel from a cheap fashion headset, functional, but imprecise.) The DXF ships alongside the STLs so anyone can order their own.

3D-print guidance

- **Material:** PETG (durable, slight flex, good for cups and the stressed gimbal/headband parts) or PLA+ (easy, rigid, fine for prototypes). Avoid brittle plain PLA on any stressed part.
- **Tolerances:** start 0.2 mm clearance for a friction fit, 0.3–0.4 mm for a slip fit; tune to your printer.
- **Heat-set inserts:** M3 brass inserts in printed bosses for any joint opened and closed (driver retention, gimbal), far more durable than screwing into plastic.
- **Orientation:** put the best surface on visible faces; print the grille flat; orient the yoke so stress runs *along* layer lines, not across them (layer adhesion is the weak axis).
- **Supports:** designed to be support-free where possible (overhangs ≤45° chamfered).

Parametric setup (for builders who want to modify it)

The model is driven by named parameters so it can be adapted without re-modeling. Starting values:

```
cup_interior_diameter = 78 mm
cup_depth             = 30 mm
wall_thickness        = 3 mm
driver_od             = 42 mm      ← update when driver confir
driver_aperture       = 35 mm      ← update when driver confir
gasket_channel_width  = 2 mm
gasket_channel_depth  = 1.5 mm
pad_lip_od            = 62 mm
pivot_post_diameter   = 6 mm
vent_slot_count       = 7
vent_open_fraction    = 0.40
```

Cup approach, interior first. Model the acoustic void as the primary geometry (revolve the half-profile of the interior, then shell outward to `wall_thickness`), then add the pad lip, baffle flange, yoke boss, and vent array. Build order: cup → baffle → yoke → slider, each constrained by the part before it.

Printing the parts without your own printer

If you don't have a printer, FDM print services (e.g. JLCPCB, Xometry) will print the part set inexpensively: order a few iterations at once to batch the shipping wait. For a show finish on the cups, an SLA service gives a smoother surface. Once a design is stable and you're iterating often, an inexpensive desktop printer pays for itself quickly.

Bill of materials (build-it-yourself)

COMPONENT	SOURCE	EST. COST
2× 40 mm driver, 32 Ω	Parts Express / Madisound	$8–18

1× Brainwavz HM5 velour pads	Brainwavz	$20-25
1× spring steel arc (laser-cut, formed)	SendCutSend + form	$8-15
8× M3 heat-set inserts	hardware/online	$1
8× M3×8 mm button-head screws	hardware/online	$1
Foam gasket tape (small roll)	hardware	$1
Damping pack (foam + felt + fiberfill)	Parts Express	$3-5
Cable + Y-split + 3.5 mm TRS plug	online	$8-15
Printed parts (2× cup, baffle, yoke, slider)	own printer or service	$3-25
Approx. total		**~$55-105**

Own-printer build lands near the low end (filament only); print-service sourcing near the high end.

Open design questions

These are the live problems still being worked, documented as the design develops.

Spring steel arc radius. Arc radius and steel thickness together set clamping force: too tight is uncomfortable, too loose falls off. Approach: print forming jigs at 95 / 105 / 115 mm inside radius, form a blank over each, test across head sizes before committing. Cold-formed spring steel springs back ~10–15% when released, so cut the jig radius 10–15% *smaller* than the target. If the best radius gives the wrong clamping force, change thickness (0.6 mm lighter, 0.8 mm firmer) before changing geometry.

Driver confirmation. The baffle is parametric specifically so the production driver can be locked late: update `driver_od` / `driver_aperture` and the baffle follows. Confirm a specific driver against the selection criteria above, ideally one with published measurements.

Locked vs. open — for the form/industrial-design work

Locked (the design must respect these):

- The four-part architecture and the project's own 6 mm pivot interface
- Driver seat parametric to `driver_od`; baffle aperture to `driver_aperture`; grille protects the driver
- Pad lip sized to the Brainwavz HM5 (61–62 mm OD)
- Circumaural fit; spring steel headband, 1095 / 0.7 mm / 10 mm
- Standard cable connector

Open (the form playground):

- Cup profile and outer form
- Rear vent geometry and pattern: the signature look
- Yoke / slider styling within the pivot and clamp constraints
- Color, finish, and overall aesthetic language

Design files license

The Daily Driver is open source under the MIT License, one license over the whole repo, design and docs alike. Download it, modify it, build it, and sell what you make, commercial use included. All we ask is credit to makerphones by Warren Labs. More on the license page.

---

v0.2 · 2026-06-12 · Reconciled from the earlier Daily Driver brief and the driver-grounded acoustic spec. Update as the form lands and the first prototype is measured. Open decisions: specific production driver, final pad confirmation, connector choice. License: MIT.

Daily Driver — Parts & Exploded View

Every Daily Driver part you print, in 3D: pull the assembly apart and put it back together, isolate any part, and see exactly what to print and how.

The Daily Driver, assembled.

> **On the web:** rotate, explode, and isolate every part in the interactive viewer at **makerphones.com/learn/daily-driver-parts**.

This is the Daily Driver as an interactive model. Drag the slider to **pull the assembly apart and back together**, toggle the **sub-assemblies** (earcup, gimbal, headband), or **click any part to isolate it** and inspect it on its own.

This page shows the **current build** — the latest version promoted as stable. To watch the design change in real time as it's worked on, see the **live beta view**. It's all developed in the open at github.com/makerphones/daily-driver.

The headphone is built from a small set of 3D-printed parts plus one bought metal headband bow. Everything below is exactly what you print to make one pair: the quantities are per complete headphone (two ears).

Parts to print

For sourcing the bought hardware (the bow, M3 shoulder screws, and heat-set inserts) and rough costs, see the build's bill of materials in the design spec. Each printed part is a starting point you can modify. The model is fully parametric, so changing a driver, a pad, or the head size regenerates the parts to match.

ⓘ CURRENT BUILD VS. LIVE BETA

This is the **current build** — the latest version promoted as stable. The live beta view tracks the work-in-progress and updates on every commit. Nothing is print-verified yet; some parts (the headband pad, the head fit/kinematics) are deliberately rough drafts. Follow along, and fork freely.

Designing Headphones with AI

A working framework for AI-assisted headphone design: where the current tools genuinely save time, where they don't, and how to fold them into the design process without building tooling instead of designing.

AI tools started saving real time on headphone design work across 2025–2026, but in narrow, specific ways, and most "design a headphone by talking to it" claims oversell what's real. This is a framework from actually doing the work: where AI saves real time, where it doesn't, and how to fold it in without getting lost building tooling instead of designing.

The core insight

AI saves you time on the work around the design. It doesn't do the design. Decades of audio domain knowledge, driver selection, acoustic intuition, tuning judgment, is the irreplaceable part. What AI takes off your plate is the mechanical work *around* the design: parametric CAD setup, concept rendering at scale, document generation, lookup and synthesis. So more of your time goes to the judgment calls that decide how the headphone sounds: driver choice, target curve, voicing.

The phase model

A headphone project moves through six phases, each with different AI utility. (The phases here are the AI-utility view of the same workflow the design methodology (Chapter 13) chapter covers as a build process.)

PHASE	WORK	AI UTILITY
1. Brief	Spec, decisions, sources	**High:** a thinking partner for articulating trade-offs

2. Acoustic design	Driver, cup volume, damping, target curve	**Low:** domain judgment dominates
3. Industrial design	Form, materials, finishes	**High:** image generation and render tools for exploration
4. Mechanical CAD	Parametric model, drawings, BOM	**Medium:** saves setup time; hands-on for surfaces
5. Prototype & tune	Build, measure, iterate	**Low:** physical and acoustic work; AI only at the documentation edges
6. Write-up	Public documentation	**High:** a writing partner; AI imagery for visuals

The pattern: AI pulls its weight on briefing, write-up, and the visually-driven middle, industrial design. The audio-engineering core, acoustic design and prototype tuning, stays a domain-expertise game.

The tools, and where they earn their keep

The tools and pricing below are a snapshot as of June 2026. This space moves fast. Verify current details before relying on any of it.

Claude (conversation). The most useful AI tool across every phase: design briefs, decision logic, parameter-table setup, BOM analysis, articulating trade-offs, cross-referencing existing products, drafting prompts for other tools. The Daily Driver design spec is a worked example of what this is genuinely good at: parametric thinking, sequencing, and supplier insight, without rendering a single piece of geometry.

Claude Code. For scripted work: orchestrating multi-step pipelines, editing many files at once, running batch operations against APIs. Where you'd reach for the command line, it's a productive layer on top.

Fusion 360 + Claude (MCP). Autodesk and Anthropic shipped a Fusion integration on April 28, 2026, as part of Anthropic's "Claude for Creative Work" launch, one of nine design-tool connectors alongside Blender, Adobe, SketchUp and others. It lets Claude take real actions inside Fusion through natural language; Autodesk structured it as two MCPs, one for creating and modifying geometry and one for querying the model. The connector is available across Claude plans, but you need a Fusion subscription to use it; set it up through the Fusion connector in Claude per Autodesk's official instructions.

- *Good at:* parameter tables, basic sketches and extrudes, repetitive geometry, unit switching, querying the model, a useful accelerator when you're rusty on Fusion.
- *Not good at:* taste-driven surfaces, complex fillets, vent-slot arrays, the parting line between cup shell and baffle, the fit-and-finish where industrial design lives. Natural language gets clunky describing what a couple of mouse clicks accomplish.
- *Bottom line:* worth setting up; expect it to handle maybe a third of the CAD work and stay out of the way for the rest.

Vizcom. Built for industrial designers: takes a hand sketch and renders it photorealistically while preserving the original line work; the Modify tool iterates materials, colors, and lighting without altering the geometry. Free tier covers ~10 renders/month at 720p with watermarks; paid is around $29/month for unlimited 4K. Fits Phase 3: rough-sketch a cup-and-yoke profile, render it in brushed aluminum, then walnut, then matte black-anodized.

Midjourney / Pinterest. Upstream mood and concept exploration: "what does this kind of headphone look like" before you have sketches to render. Pinterest does the same job with human-curated breadth instead of AI-generated variation.

FAL.ai. Serverless API access to current image-generation models (Flux, SDXL, and others), pay-per-call, scriptable. Useful for higher-volume concept exploration than Midjourney with more API control, and for hero imagery in the write-up, especially once you've trained a style model (LoRA) for consistent output across a product line.

Image-to-3D (Meshy, Trellis, Hunyuan3D). Convert a 2D render to a rough 3D mesh. Quality varies and the output isn't manufacturing-ready. Possible use: turn a chosen render into rough geometry as a reference body to import into CAD, then rebuild it parametrically. Worth knowing about; not a primary path.

OpenSCAD and CadQuery (code-based CAD). Code-based parametric CAD libraries. Claude can write the script, you run it locally, you get an STL, no MCP needed, because the workflow doesn't require one. Less polished than Fusion (no surface tools, no fillet UI), but the iteration loop is dead simple. For open, reproducible designs, the code-based path has real merit: the files are text, they version-control naturally, and anyone can modify them.

Physical / production. SendCutSend for laser-cut spring-steel arcs from a flat DXF (developed-arc geometry; ~$8–15 per piece at low quantity, dropping at volume); FDM print services (JLCPCB: cheap, slow) and SLA (Shapeways or Xometry: quality, expensive), or an inexpensive desktop printer once iteration frequency justifies it; Brainwavz HM5 pads as a common pad target for DIY headphones.

Three architecture patterns: MCP, RAG, and APIs

Three patterns, often combined.

MCP (Model Context Protocol): a standard for connecting AI models to external tools and services so the model can act on them (Fusion's MCP, and many more arriving). About *doing things* in other systems through the model.

RAG (Retrieval-Augmented Generation): the model pulls relevant context from a knowledge base, usually vector-searched over your documents. About *grounding the model* in your own information.

API calls from scripts: batch work where the model isn't in the loop turn-by-turn: image generation, image-to-3D conversion, documentation rendering.

For a headphone project the working combination is: a **knowledge base** of your chapters, specs, and measurement data for grounding; **interactive tools** like Fusion during CAD work for action; **batch services** for image generation and image-to-3D when you need volume; and plain **file edits** for documentation. That's it: a knowledge base, the tools you're already in, and some batch scripts.

Principles that keep you out of trouble

1. Build the pipeline as the project needs it. The most common trap is building elaborate AI infrastructure before doing any actual design work. Each tool earns its place when the phase that uses it begins; a parked backlog, evaluated at phase start, is the antidote.

2. Document everything as you go. For open designs, the documentation *is* the product: every decision, every measurement, every iteration. Build the documentation habit into the workflow from day one. AI is a real help with the documentation work itself: drafting, restructuring, keeping decision logs current.

3. Phase-gate tool adoption. Don't evaluate a render tool in Phase 2. Don't set up Fusion MCP in Phase 1. Wait until the phase needs it. This protects focus more than any productivity hack.

4. Keep the judgment calls in your hands. Driver selection, acoustic intuition, tuning preference, voicing decisions. These stay with the designer. AI that pretends otherwise is overselling.

5. The hard parts stay hard. Acoustic design takes time. Prototype iteration is unpredictable. Tuning is a months-long process. AI doesn't shortcut any of it. It just makes the surrounding work less of a tax.

6. Two CAD paths, choose deliberately. Fusion 360 + Claude MCP for polished work where industrial-design fit-and-finish matters. OpenSCAD or CadQuery for code-first, reproducibility-first work, which, for open and remixable designs, has real merit.

Tool details above are a June 2026 snapshot and will drift. Verify before relying on them. The framework is the durable part.

REFERENCE

Appendices

APPENDIX

Listening Safely

Building headphones means putting transducers an inch from your eardrums and then, during measurement and tuning, running loud test tones through them over and over. That's a combination worth respecting. Your ears are the one instrument in this whole hobby you can't replace, recalibrate, or print a new copy of, so a few simple habits are worth building in from the start.

The two real risks

Your hearing is at risk here in two ways. The first is a sudden loud blast from a fault: a source left at full volume, a driver fed a signal you didn't expect, a glitch, or just plugging in while the volume's up. A spike like that, right against your ear, can do damage in an instant. The second is slower: long tuning sessions where the volume creeps up without you noticing, and an hour later you're listening far louder than you started.

The habits that handle it

The fixes are easy, and worth making automatic.

Turn the volume all the way down before you plug anything in or start a new sweep, then bring it up gradually. This one habit heads off most of the sudden-blast scenarios on its own.

Don't wear an untested build while you first power it up or run a new test signal. Hold it off your ears, or start it whisper-quiet, and confirm nothing's wrong before it's sealed against your head.

Keep test levels modest. You don't need a loud sweep to get a clean measurement. Moderate is plenty, and it's gentler on both your ears and the driver.

Watch the clock and the creep. If you've been tuning for an hour and keep nudging the volume up, that's your ears getting tired. The headphones haven't changed – you have. Take a break. Your read on the sound gets worse as your ears tire anyway, so breaks make you a better tuner, not just a safer one.

A rough yardstick

Around 85 dB is roughly the level you can listen to all day safely, and the safe time drops fast above that: cut about in half for every 3 dB louder. The catch is you can't really know your level by ear, which is exactly why the habits matter more than any number. A simple gut check: if the music's loud enough that you couldn't hold a normal conversation over it, it's loud enough to add up over time.

None of this is about being precious. Hearing damage is cumulative and permanent, and tinnitus is a genuine occupational hazard in audio. Plenty of people who've spent a career around sound have the ringing to show for it. Protecting your ears isn't caution for its own sake. You're protecting the only tool you've got for judging whether the thing sounds good.

APPENDIX

Glossary

The manual uses a fair amount of vocabulary, and it's all defined where it first comes up, but here it is collected in one place to look up as you go.

AAC: A Bluetooth audio codec that performs well on Apple devices; one of several codecs a source and headphone may share.

aptX: A family of Qualcomm Bluetooth codecs (aptX, aptX HD, aptX Adaptive, aptX Lossless) offering better quality than the SBC baseline, provided both ends support it.

Baffle: The plate the driver mounts to, separating the front of the diaphragm from the rear. The driver-to-baffle joint is a critical seal.

ABS: Acrylonitrile butadiene styrene, a common engineering thermoplastic with good mechanical properties and temperature resistance. Difficult to FDM print due to warping, but vapor-smoothable with acetone.

Balanced armature (BA): A small, efficient driver type that covers a limited frequency band, common in in-ear monitors where several are combined with a crossover.

Bass reflex: An enclosure design that uses a tuned port to extend bass response below the driver's sealed-box rolloff. Also called a vented enclosure.

Break-up mode: A condition where different parts of the diaphragm move non-uniformly, producing resonances typically above 5 kHz.

Boundary element method (BEM): A simulation technique that solves the wave equation across a 3D mesh, used for radiation and diffraction. More accurate and more demanding than lumped-element modeling.

Channel matching: How closely the left and right channels agree. Poor matching smears the stereo image; it's both a mechanical and an electrical concern.

Closed-back: A design with a sealed rear chamber, giving isolation and stronger low bass at the cost of a more closed-in sound.

CLD (Constrained Layer Damping): A damping technique using a viscoelastic material (like Dynamat) bonded to a vibrating structure to convert vibration to heat.

Codec: The compression scheme used to send audio over Bluetooth (SBC, AAC, aptX, LDAC, LHDC, LC3). Both the source and the headphone must support a codec to use it.

Coupler: An artificial ear canal with a microphone, used in a measurement rig. The IEC 60318-4 standard defines the common type.

Compliance: A measure of springiness, how easily something (a suspension, a volume of trapped air) yields to force. Larger compliance is a softer spring.

CSD (cumulative spectral decay): A "waterfall" plot showing how energy at each frequency decays over time, revealing resonances that ring on after the signal stops.

Damping: Material (felt, foam, fiber fill) added to absorb acoustic energy, used to tame resonances, reflections, and excess bass.

Diaphragm: The thin membrane that moves to push air and create sound. The heart of any driver.

Diffraction: The bending of sound waves around obstacles (like a baffle edge), creating interference patterns that affect frequency response.

Driver: The transducer that turns an electrical signal into sound. Dynamic, planar magnetic, balanced armature, and electrostatic are the main types.

Dynamat: Brand name for constrained layer damping material (butyl rubber plus aluminum foil), applied to cup interiors to damp structural vibrations. It is not acoustic absorption.

Dynamic driver: The most common driver type: a voice coil attached to a diaphragm, moving in a magnetic gap. Simple, hard to break, good bass.

Electrostatic: A driver type using a charged film between electrodes; very low distortion, but requiring special high-voltage amplification.

Frequency response (FR): How loud a headphone plays at each frequency, the single most-used measurement. Usually shown as a curve.

Front chamber: The acoustic space between the driver face and the ear. Its volume is set primarily by ear pad depth, and it affects high-frequency response and soundstage character.

Fs: A driver's free-air resonance frequency, one of its Thiele-Small parameters.

Group delay: The time delay experienced by each frequency in a system. Non-uniform group delay means different frequencies arrive at the ear at different times.

Harman target: A research-based preference curve describing how a neutral-sounding headphone tends to measure on a standard rig. A common reference target, not a law.

IEC 60318-4: The standard (often called "711") for the coupler used in headphone and IEM measurement. Reliable to roughly 10 kHz.

Impedance: The driver's opposition to current, in ohms, and how it varies with frequency. Affects how it pairs with an amplifier.

Impulse response: A measurement showing how a headphone responds to an instantaneous transient. Reveals ringing, resonances, and time-domain behavior not visible in frequency response alone.

LC3: The codec for Bluetooth LE Audio, more efficient than older codecs and the basis for broadcast features like Auracast.

LDAC: Sony's high-bitrate Bluetooth codec aimed at high-resolution audio.

LiPo (Lithium Polymer): A rechargeable battery chemistry used in wireless audio devices, with a nominal voltage of 3.7 V. Requires a dedicated charging circuit and protection against over-discharge.

Lumped-element method (LEM): Modeling a driver-and-chamber system as an equivalent electrical circuit of masses, compliances, and resistances. The tractable approach for headphone simulation.

Mass loading: Adding mass to a vibrating structure to lower its resonant frequency and damp structural vibrations. Different from acoustic damping: it affects the structure, not the air.

Nominal impedance: A single representative impedance value for a driver, typically measured at 1 kHz. The actual impedance varies with frequency.

Open-back: A design with a vented or open rear, giving a more spacious, natural sound but little isolation and bass that leaks out both ways.

PETG (Polyethylene Terephthalate Glycol): A 3D printing filament with better impact resistance and temperature tolerance than PLA. Recommended for structural headphone components.

Pinna: The fleshy outer ear. In a measurement rig, a silicone pinna sits in front of the coupler to load over-ear headphones realistically.

PLA (Polylactic Acid): The most common 3D printing filament. Easy to print but brittle under repeated stress, with a low glass transition temperature (~60°C). Good for prototyping; switch to PETG for final builds.

Planar magnetic: A driver type using a flat diaphragm with an embedded conductor between magnet arrays; even, low-distortion sound, usually heavier and less efficient.

Polarity: Which way a driver moves for a given signal. Both channels must be in phase, or the bass cancels and the image collapses.

PPI (Pores Per Inch): A specification for acoustic foam density. Higher PPI means smaller pores and more absorption of higher frequencies. Typical range for headphone use is 20–60 PPI.

Q factor (Qts, Qtc): A dimensionless parameter describing the damping of a resonance. High Q means a sharp, peaked resonance; low Q means a broad, well-damped one. Qts is for the driver in free air; Qtc is for the driver in a closed enclosure.

Rear chamber: The enclosed volume behind the driver in a closed-back headphone. Its volume directly affects bass extension and resonance behavior.

Resonance: A frequency at which something naturally wants to vibrate, where energy piles up and rings. Sources include the driver, the air cavities, and the structure.

REW (Room EQ Wizard): The free, de facto standard measurement software used across the hobby.

SBC: The baseline Bluetooth audio codec that every device supports; the guaranteed fallback when nothing better is shared.

Sealed: Fully closed, with no leak. A sealed rear chamber maximizes isolation and bass but can produce a resonance peak.

Sensitivity: How loud a driver plays for a given input, often quoted in dB per milliwatt or per volt. Together with impedance, it tells you how easy a headphone is to drive.

Sorbothane: A proprietary viscoelastic material used for vibration isolation and damping. Effective for reducing the mechanical transmission of vibrations.

Soundstage: The perceived sense of space and width in the sound, influenced strongly by open versus closed design.

SPL (sound pressure level): Loudness, measured in decibels. Calibrating SPL lets a measurement read absolute levels.

Standing wave: A resonance in a cavity where a reflected wave reinforces itself at a particular frequency, building a fixed pattern of loud and quiet regions.

Thiele-Small parameters: A driver's key electro-mechanical numbers (resonance, compliance, moving mass, motor strength, and more) used to model and predict its behavior.

THD (total harmonic distortion): How much a driver adds harmonics not in the original signal. Usually low; a sudden spike often signals a mechanical fault.

TP4056: A widely used single-cell LiPo battery charging IC, available as inexpensive module boards. The standard DIY choice for Bluetooth headphone charging circuits.

TRRS: A four-conductor plug (tip, ring, ring, sleeve) carrying left, right, ground, and mic, the single-jack headset standard. The CTIA pinout is the modern standard.

TRS: A three-conductor plug (tip, ring, sleeve) carrying left, right, and ground, the standard stereo audio connector, with no microphone contact.

Vas (equivalent compliance volume): The volume of air with the same acoustic compliance as the driver's suspension. A larger Vas means more compliance and benefits from a larger enclosure.

Voice coil: The coil of wire attached to a dynamic driver's diaphragm; current through it in the magnetic gap is what moves the diaphragm.

APPENDIX

Supplier Directory

Where to source what, organized by what you're looking for. A word of warning before anything else: suppliers and prices change faster than any other information in this manual. Shops close, stock vanishes, models get discontinued. This very directory will go stale. So it leans on durable, established sources and on sourcing *strategies* that outlast any single shop, and the most reliable move is always to check the community for what people are actually buying right now.

Drivers

The driver is the heart of the build, and there are a few durable routes to one.

Dedicated DIY-audio retailers are the backbone. In the US, Parts Express has supplied DIY audio for decades and carries raw drivers (its Dayton Audio line among them) along with components and tools. The broad marketplace, AliExpress and similar, offers raw dynamic and planar drivers in enormous variety at low prices. The catch is that quality varies wildly, so cross-check anything against community experience before committing. Some headphone makers sell replacement drivers as spares, which is another path to a known-good transducer. And for the truly from-scratch builder, micro-speaker and component suppliers exist, though they're aimed more at the electronics industry than at audio hobbyists.

Don't overlook salvage. An inexpensive or broken donor headphone is a perfectly good driver source, and plenty of excellent builds start with a salvaged or cheap driver. Driver selection guide (Chapter 7) covers what to look for whichever route you take.

The modding-platform route

If you'd rather learn by modifying than by building from a bare driver, there's a legendary on-ramp: the Fostex T50RP. Now in its fourth generation, it's a stock planar-magnetic headphone practically designed to be opened up, damped, and re-voiced, with a famously improvable stock sound and one of the deepest modding communities in the hobby. A whole ecosystem surrounds it, from community designs like ModHouse's Argon to ready-made mod kits and sanctioned pad-and-driver variants.

Modding a T50RP is many builders' first serious project, and it teaches the same fundamentals as a ground-up build (chambers, damping, seal, measurement) with the structure and the driver already sorted. It's a genuinely good place to start.

Ear pads

Pads are acoustic parts as much as comfort parts, so this is a real decision. Dekoni Audio, an established American maker, and Brainwavz are the well-known aftermarket pad sources, with options across sizes, shapes, and materials. Manufacturer OEM pads and generic marketplace pads round out the field. Remember that swapping pads re-voices a headphone, see ear pads and comfort (Chapter 10), so treat a pad change as a tuning decision rather than a comfort upgrade.

Damping materials

You rarely need a branded kit. Acoustic foam, felt, fiber fill or natural wool, and even modeling clay for baffles are inexpensive from general craft, hardware, and online sources, and many builders buy what amounts to a lifetime's supply for the price of a single branded mod kit. The branded kits save you the cutting and the guesswork, which has value for a first project, but the materials themselves are commodity items. Damping materials (Chapter 11) covers what each one does.

Cable and connectors

DIY-audio retailers and the marketplace carry headphone cable and the connectors you'll need: 3.5 mm and 2.5 mm plugs, mini-XLR for balanced connections, and MMCX or two-pin for in-ear monitors. Cables, connectors, and hardware (Chapter 12) covers the choices and the wiring.

Filament and 3D printing

Standard PLA and PETG from any filament supplier will cover most builds: PLA for ease, PETG where you want a little more toughness and damping. If you don't own a printer, online print services and local makerspaces will print your parts from a model, which is a low-commitment way to start before investing in a machine of your own.

Measurement gear

This is covered in detail in budget measurement setup (Chapter 19), so in brief: miniDSP sells the UMIK measurement microphones and the EARS and EARS Pro headphone fixtures directly; calibrated-microphone specialists offer individually corrected UMIKs; and the broad marketplace is where IEC 711 clone couplers live, alongside the printable measurement stands shared by the community.

Electronics for wireless and active builds

For Bluetooth and other powered builds, specialist audio-electronics retailers and the marketplace carry the pieces: Bluetooth receiver boards built around the common ng and protection boards. Bluetooth integration `(Chapter 30)` covers what to look for, especially matching a board's codec support to the devices you actually use.

A standing caution

Treat named shops as durable starting points rather than permanent fixtures, lean on salvage and the broad marketplace when you need flexibility, and check the community for current recommendations before any significant purchase. The strategy lasts longer than any shop on this list.

APPENDIX

Design Resources

A short, curated set of the tools and communities that make building easier. This covers software, model libraries, and people, not parts suppliers, which have their own directory. Tools change, so treat this as a starting map rather than a fixed list, and favor whatever you'll actually learn well over whatever is newest.

CAD and 3D modeling

For designing cups, baffles, and fixtures, you want parametric CAD: the kind where dimensions are editable and the model updates. Common choices range from free, capable options for hobbyists to subscription tools with generous free tiers for personal use. For the organic shapes of pads and ergonomic surfaces, a mesh-and-sculpt tool complements parametric CAD. Pick one and get fluent in it; switching tools costs more than any feature gains you. See 3D design for headphones (Chapter 14) for how to approach the modeling itself.

Slicing and printing

Your slicer turns a model into printer instructions, and most printers ship with a recommended one, start there. What matters more than the slicer is a calibrated, documented print profile you reuse, so your parts come out the same every time, which is the foundation of manufacturing for consistency (Chapter 28). PLA is the easy starting material; PETG adds a little toughness and damping for structural parts.

Measurement

Room EQ Wizard (REW) is free and the de facto standard for acoustic measurement across the hobby: it handles the sweep, the calibration files, smoothing, decay plots, and overlays. Pair it with a calibrated measurement microphone or an IEC 711 clone coupler, as described in budget measurement setup (Chapter 19), and you have everything you need to measure and tune.

Two free tools sit alongside REW once you have a measurement in hand. **AutoEQ** (github.com/jaakkopasanen/AutoEq) is an open-source project that generates target-based EQ profiles from measurement databases: it carries a large library of commercial headphone measurements, and it's useful both as a reference for understanding target-curve deviations and as a starting point for tuning your own builds toward an established target. **EqualizerAPO + Peace** is Windows-based system-wide EQ, primarily for listening, but it's handy because it lets you test EQ corrections derived from measurements by ear before committing them to hardware tuning: fast A/B comparison of different compensation curves, no soldering required.

Acoustic modeling and simulation

For predicting how a driver and chamber will behave before you build, the accessible tools work on the lumped-element approach. **Hornresp** is free and widely used and handles sealed and vented systems well. **AKABAK** is a professional electroacoustic simulator with a free version that does everything but save results, coupling lumped-element and boundary-element methods. **VituixCAD** is free and excellent for crossovers and diffraction, more loudspeaker-oriented but useful background. Full finite-element packages exist but are priced for industry. Acoustic modeling (Chapter 26) covers how to use these and, importantly, their limits.

Model libraries

Public 3D-model repositories are full of headphone cups, measurement stands, and fixtures shared by other builders, including printable measurement rigs designed around common clone couplers. They're an excellent way to start from a known design and modify rather than beginning from a blank sketch, especially for a first build. Always check that a shared design is sized for your specific driver before printing.

Measurement databases

Community measurement databases let you browse and compare frequency-response graphs for a huge range of headphones and IEMs. They're a useful reference for what target you might aim at and how commercial designs measure, but remember that different contributors use different rigs, so the graphs aren't strictly comparable to each other or to your own. Treat them as a rough guide to target and overall shape, not numbers you can hold your own measurements against. The major databases, and the preference research and standards behind them, are mapped in Sources & Further Reading.

Communities

The DIY headphone and audio communities are where the real knowledge lives, and where you'll get answers to the specific questions this manual can't anticipate. Long-running forums cover both headphones broadly and DIY audio specifically, with deep archives worth searching before asking. There are active discussion communities on the major social platforms as well. Bring your builds, your measurements, and your failures: the failures often spark the most useful conversations, and learning in public is how everyone here got good.

A note on staying current

Software versions, module availability, and prices move faster than a reference manual can. When a chapter names a specific tool or part, treat it as a pointer to a category as much as a recommendation, and check current options before you buy. The principles hold; the products around them won't.

APPENDIX

Sources & Further Reading

Reliable information on headphone design is scattered, and it clusters at two ends: deep acoustic theory, and a fast-moving pile of measurement data, with surprisingly little practical "how to actually build one" in between. That gap is much of why Makerphones exists. This page is the shelf we wish we'd had: the credible sources worth your time, sorted by what they're actually for, with an honest note on where to start.

Where to start

If you're new and don't want to be dropped into graduate-level acoustics cold:

- **Read Sean Olive's free summary of the Harman headphone research first** (in *Acoustics Today*, 2022, see below). It's free, about 6,000 words, and it's the single best on-ramp to how modern headphone sound is understood.
- **Look at real measurements** on Crinacle's or oratory1990's databases until the frequency- response graphs stop looking like noise. Pair that with the manual's own chapter on reading frequency response (Chapter 2).
- **Then,** if you want the *why* underneath it, Floyd Toole's book for the psychoacoustics and Beranek & Mellow for the hard theory.

The rest of this page is the full shelf.

Foundational theory — acoustics and transducers

The bedrock. Rigorous, math-heavy, and not light reading, but the physics actually lives here.

- **Beranek & Mellow, *Acoustics: Sound Fields and Transducers*** (and the second edition, *Acoustics: Sound Fields, Transducers and Vibration*). Tim Mellow's modern update of Leo Beranek's classic 1954 *Acoustics*. Covers earphones and transducers through electro-mechano-acoustical circuits, with worked examples aimed at practical design. The reference for the underlying physics.
- ***Loudspeaker and Headphone Handbook*, edited by John Borwick** (3rd ed., 2001). Mostly a loudspeaker book, but **the "Headphones" chapter (ch. 14, by C.A. Poldy)** is an extended, genuinely comprehensive treatment of headphone acoustics, one of the most thorough single references on the subject in print.

The science of "good" — psychoacoustics and preference

The modern revolution: rigorous listening research into what people actually prefer, which gave the field a shared target.

- **Floyd Toole, *Sound Reproduction: The Acoustics and Psychoacoustics of Loudspeakers, Rooms and Headphones.*** The definitive text on the perception side of reproduction; recent editions fold in headphones. Essential for understanding *why* measurements matter and how they connect to what you hear.
- **The Harman headphone research (Sean Olive, Todd Welti, and colleagues).** A roughly nine-year listener-preference study that produced the Harman Target Curve, reported across 19-plus AES papers. The most influential body of modern headphone science. Access points:
    - *Free starting point:* **Olive, "The Perception and Measurement of Headphone Sound Quality: What Do Listeners Prefer?"**, *Acoustics Today* (2022), a free PDF distilling the research in plain language.
    - *Accessible writing:* **"Audio Musings by Sean Olive"** (seanolive.blogspot.com).
    - *Key papers:* the "A Statistical Model that Predicts Listeners' Preference Ratings of... Headphones" series (in-ear, then around-ear and on-ear), AES conventions, 2017–2018.

Foundational papers, and where the literature lives

- **Møller, Jensen, Hammershøi & Sörensen, "Design Criteria for Headphones,"** *Journal of the Audio Engineering Society*, Vol. 43, pp. 218–232 (1995). An early, foundational paper on the problem of designing headphones to a defined acoustic goal.
- **The AES E-Library (aes.org).** The canonical archive of audio-engineering papers, including most serious headphone research. Paywalled, but it's where the primary literature lives, and AES membership or a university library gets you in.

Standards

How measurement gets defined, so different rigs can agree.

- **IEC 60318 series:** specifications for ear and head simulators (the "ear simulator," and the widely-used occluded-ear/"711-type" coupler) that measurement rigs are built to.
- **ISO 11904:** methods for determining sound immission from sources close to the ear (the diffuse-field and free-field references).
- *Worth knowing:* the older diffuse-field calibration that these standards recommend is now widely regarded as outdated, having been largely displaced by

rence-based targets like the Harman curve.

Measurement data and the modern ecosystem

Where makers and enthusiasts actually live now: public databases and open tools built on the research above.

These are live sites, tools, and hardware. Links, ownership, and methods change. Treat the specifics as current-ish and verify before relying on them.

- **oratory1990:** measurements on the gold-standard GRAS 45BC rig, referenced to the Harman target; widely treated as the reference source for over-ear data.
- **Crinacle** (graph.hangout.audio): one of the largest public databases of headphone and earphone measurements, especially strong on IEMs.
- **squig.link / Squiglink:** a family of frequency-response database and comparison tools.
- **AutoEq** (open source): automatic headphone EQ generated from frequency-response data, aggregating measurements from oratory1990, Crinacle, and others; a good way to see how measurements, targets, and corrections connect.
- **Others:** Rtings, Reference Audio Analyzer, and SoundStage Solo (Brent Butterworth), among more. The underlying hardware is usually a GRAS rig (the 45-series, or the 43-series).
- **A standing caveat:** measurements shift with the rig, the seating, and the pad wear, and a headphone never sounds *exactly* like its graph. Read measurements as strong guidance, not gospel, which is why good measurement means averaging several seatings (Chapter 20).

Communities

Where the conversation happens and questions get answered.

- **Head-Fi:** the largest and longest-running headphone enthusiast community.
- **Audio Science Review (ASR):** measurement- and data-focused discussion.
- **r/headphones, the DIY-audio forums, and the Fostex T50RP modding community:** where the practical build-and-mod conversation actually happens, scattered as it is.
- **The Audio Engineering Society (AES):** the professional body, its conventions, and its E-Library; the front door to the formal literature.

This list leans toward theory and measurement because that's what credible, lasting sources exist for. The practical bridge between them, actually designing and building a headphone, is what the rest of Makerphones is for. Found a source worth adding? There's a feedback link on every page.

APPENDIX

Troubleshooting Guide

When a build sounds wrong, the cause is usually one of a handful of familiar culprits. This guide is organized by what you're hearing, so you can jump to your symptom, find the likely cause, and fix it. Each points back to the chapter that goes deeper.

No bass, or thin and hollow sound

Almost always one of two things. First, **reversed polarity**: if one channel is wired out of phase, the two cancel in the bass and the stereo image goes vague and hollow. Play a mono track; it should image as a single point dead center. If it sounds wide or sucked-out, swap the leads on one channel. Second, a **broken seal**: air leaking around the driver-to-baffle joint or between the cup and your head collapses the low end. Press the cups gently harder; if the bass swells, you have a leak to chase. See driver mounting and assembly (Chapter 16).

Boomy, one-note, or boxy bass

This one lives in the chamber. A **rear chamber that's too small** is the classic cause: the trapped air stiffens and piles the bass onto one frequency. The fix is volume or venting, not foam: open a vent, or move to a larger rear volume. A small, pluggable vent opened one hole at a time is the cheapest way to dial it in. See acoustic chamber design (Chapter 15).

One channel dead or quiet

A wiring fault. Check the **solder joints** at the driver and the connector: a cracked joint at the driver is the single most common long-term failure, usually from cable flex with no strain relief. A multimeter helps you find the break. Re-solder, and add strain relief at both ends so it doesn't happen again.

Left and right sound different

A **channel mismatch**, and it's mechanical as often as electrical. Confirm both cups have the same damping (weigh the fill), the same seal, the same pads, and matched drivers. The reliable cure is to rebuild both channels identically, in one session, and a quick measurement overlay makes a mismatch you can hear but can't place obvious. See manufacturing for consistency (Chapter 28).

Harsh, hard, or fatiguing treble

Usually a **resonance** ringing somewhere. If it's a sharp peak, a thin ring of damping in the front cavity can tame it, but apply it sparingly, because too much dulls the whole top end. If the hardness is more of a smear, it's often the cup walls or reflections; line the walls with felt. See damping strategy and application `(Chapter 17)`.

Dead, lifeless, no air

You've **over-damped it.** The beginner instinct is to stuff the cup full "to be safe," and it drains the life out of the sound. Pull material back out until the sparkle returns. You want the smallest amount of damping that solves the actual problem, and no more.

A buzz or rattle

This is mechanical, so put the graph away. Something is loose: a **slack lead** touching the diaphragm, a part not seated, a bit of print debris in the cup, or the **driver being pushed past its limits.** With the cup open, play a sweep and listen for where it buzzes, then find and secure the loose element. On a measurement rig, a rattle shows up as a spike in the distortion plot. See advanced measurement topics `(Chapter 25)`.

The plastic cup itself rings

A **structural resonance**: thin printed walls ring like a drum, and no amount of felt fixes it, because the energy is in the plastic, not the air. Tap the empty cup: a sharp "ting" confirms it. You can't absorb your way out of this one; you have to stiffen the cup. Thicker walls, internal ribs, more infill, or a less ringy plastic. See resonance control `(Chapter 27)`.

It measures fine but sounds wrong

You've been **tuning to the graph instead of your ears.** A technically flat curve isn't the goal, and a "correct" measurement can still sound off. Trust your ears for the final verdict and use the graph to understand what's happening, not as a score to maximize. See tuning with damping `(Chapter 22)`.

The bass changes every time you put them on

That's **seating and seal variation**, and it's normal to a degree, but if it's dramatic, your seal is marginal. Check the pad-to-head seal and the driver-to-baffle seal. On a measurement rig, this shows up as scatter between re-seatings; establish a repeatable seal before you trust any measurement.

Measurements that don't match anyone else's

Expected, and not a fault. Different rigs aren't comparable, and a hobby rig isn't an industry-standard head: it's reliable to roughly 10 kHz and reads relative, not absolute. Use your measurements to track *your* changes and match *your* channels, not to chase someone else's graph. See budget measurement setup (Chapter 19).

When in doubt

Check the boring things first, in this order: is the seal good, is the polarity right, are the pads fresh and matched, is the fit right on your head. The large majority of mystery problems are one of those four, and chasing them in the tuning, when the real cause is a leak or a worn pad, is how you lose a weekend to the wrong problem.

APPENDIX

Community Builds

The rest of this manual is me telling you how to build a headphone. This section is for everybody else's builds, yours included. It's where people show what they actually made, and over time it becomes the most useful part of the whole resource, because a real build that someone documented well teaches more than any chapter can.

Why builds are worth sharing

Theory gets you started, but builds are where the hobby actually lives. A headphone someone designed, printed, tuned, and wrote up, with the choices they made and the problems they hit, is worth more to the next builder than any amount of general guidance, because it's concrete and it's proven. Someone built that, and it works, and here's exactly how.

The failures and the iterations matter just as much as the finished result. The build that sounded boxy until the rear chamber got bigger, the channel imbalance that turned out to be a gasket, the treble peak that a thin ring of felt finally tamed: those stories are how everyone learns to diagnose their own builds. Learning in public is how this community got good, and sharing what didn't work the first time is part of the deal.

What makes a build worth documenting

A great shared build is one another person could actually replicate, and that comes down to documentation. The single most valuable thing you can include is the build spec, in the spirit of manufacturing for consistency (Chapter 28): the driver you used, the chamber design and dimensions, the damping recipe by weight and placement, the print profile, and the assembly steps. "Some felt in the back" tells the next person nothing. "This much fill behind the driver, this lining on the walls" lets them build it.

Beyond the spec, a few things lift a write-up. Measurements, if you have them, so others can see what you achieved and aim at it. The process, including what went wrong and how you fixed it, because that's where the real teaching happens. And photos, because seeing the build is half of understanding it. None of this needs to be polished. An honest, well-documented build beats a beautiful one with no details every time.

Modding a commercial headphone first

Plenty of good builds don't start from a printed shell at all. They start by modifying a commercial headphone. You begin with working mechanical hardware and a driver that already sounds decent, then tune from there. The platforms below come up again and again because they're widely available, cheap enough to experiment on, and backed by communities that have already documented what each change does.

PLATFORM	COST (STOCK)	DIFFICULTY	MOD NOTES
Koss Porta Pro (on-ear, open)	~$30	Beginner	The canonical first mod. Foam replacement (the stock foams color the sound a lot), cable replacement (the stock cable fails eventually), and pad swaps. Modded, measured, and documented more thoroughly than almost any other headphone. Start here to learn what a foam change does to a frequency response.
Sony MDR-7506 (over-ear, closed)	~$100	Easy	The industry workhorse with a well-documented signature (forward upper mids, slightly rolled bass). Ear pad replacement changes front-chamber geometry and shifts bass and lower mids; driver damping to tame the 8-10kHz peak; cable replacement. A well-measured baseline to work from.
Beyerdynamic DT770 / DT990 (over-ear, closed/open)	~$180	Easy-Mid	Good drivers, metal-reinforced build. DT770 for closed designs, DT990 for open. Beyerdynamic sells replacement parts directly, including drivers, so you can build a DT-style headphone from known-quality parts. The treble peak around 8-10kHz responds well to damping.
Sennheiser HD58X / HD6XX (over-ear, open)	~$150-220	Easy	The Drop (formerly Massdrop) variants of classic Sennheiser designs: excellent drivers and build, obsessively measured. Worth modding for headband comfort and cosmetics more than sound; the acoustic design is hard to improve without replacing the drivers. Better as a reference for your own designs than a mod platform.

Grado SR80 / SR325 (on-ear, open)	~$100-300	Mid	A cult favorite in the DIY community. Good driver, deliberately minimal (some call it underdamped) housing, and an extensive mod space: cup materials, chamber volumes, damping approaches all thoroughly explored. The community has built everything from wood cups to 3D-printed alternatives.

TIP

If you've never modded a headphone, start with the Koss Porta Pro. The community documentation is unmatched, the stakes are low, and the acoustic effect of each change is well understood from years of community measurement.

What makes a good first build

The best first build is the one you actually finish. That means constrained scope, parts you can get, and a design that doesn't depend on getting everything right on the first try. A few principles make that more likely.

Use a driver with published specs. A driver with full published Thiele-Small parameters lets you check your design intent against real numbers before you cut or print anything, instead of guessing and hoping.

Plan for disassembly. You will open the headphone at least five times during tuning, so design the closure to be reopened: screws rather than glue for driver mounting, a solid mechanical pad attachment, accessible wiring. Builds that are hard to take apart stop getting tuned, which is the same as stopping before they're good.

Measure at each step. Without measurement you're guessing; with it you're iterating. You don't need expensive gear. The budget measurement setup guide covers a rig that gets you there for not much money.

Keep a dated build journal. Note what you changed, when, and what it did. Acoustic tuning has enough variables that memory alone isn't reliable, and even a plain text file with dated notes means you can reproduce your successes and actually understand your failures.

How to share yours

Bring your build to the community. Post it where builders gather, document it the way you'd want to find it documented, and answer the questions it provokes. The submissions that help most are the ones explained clearly enough that someone else can follow the path you took. Doesn't matter how fancy the headphone is.

If you're not sure your build is "good enough" to share, it is. A first build, honestly documented, helps the next first-time builder more than a tenth build presented as a finished product. Share the one you actually made, including the parts you'd do differently next time.

Add yours

A resource like this grows from its builders. Every shared failure saves the next person the same wasted evening. None of it works without builders putting their stuff up, so put yours up.

Submit your build

Built something from the manual? Send it in: **email a few photos, your parts list, and what you'd do differently to jamey@jameywarren.com** and it'll be considered for this page. (A proper submission form and community space are planned. Until then, email is the queue.)

Index

Numbers refer to chapters.

I

L

N

O

P

Q

R

S

T

V

About the author

Jamey Warren has spent more than twenty-five years in professional audio and the headphone industry. He was employee #1 at Grace Design (1997–2001) and consulted for the company through 2003 — seven years he calls his University of Rock & Roll. From 2003 to 2017 he was VP of Operations and then President & CEO of HeadRoom, where he relaunched the entire headphone-amplifier line and oversaw the testing of thousands of headphones.

He is now designing his own open-back headphone, the Daily Driver, in the open — with AI as a design partner — using the same methods this book teaches. The work, and its measurements, are published at makerphones.com.

Colophon

The Art and Science of Headphone Design was generated from the Makerphones reference manual, an open Astro + Starlight site, by a single build step that reuses the same content, diagrams, and design tokens as the website — one source, many outputs.

The text is set in **Source Serif 4**; headings in **Schibsted Grotesk**; technical labels, captions, and folios in **JetBrains Mono**. Every figure is a hand-built vector diagram, drawn for this manual.

Warren Labs · makerphones.com

www.ingramcontent.com/pod-product-compliance
Lightning Source LLC
LaVergne TN
LVHW070934210926
848482LV00003B/239
* 9 7 9 8 9 9 6 8 2 9 9 0 3 *